JOHN DALTON
1766-1844

A bibliography of works by and about him
with an annotated list of his
surviving apparatus and personal effects

A.L. SMYTH

MBE, FLA
Curator Manchester Literary and Philosophical Society

Manchester Literary and
Philosophical Publications Ltd

in association with

Ashgate
Aldershot • Brookfield USA **• Singapore • Sydney**

Published by
Ashgate Publishing Limited
Gower House, Croft Road
Aldershot, Hampshire GU11 3HR
Great Britain

Ashgate Publishing Company
Old Post Road
Brookfield, Vermont 05036–9704
USA

in association with
Manchester Literary and Philosophical Publications Ltd.
Churchgate House
56 Oxford Street
Manchester M60 7HJ

ISBN 1 85928 438 8

British Library Cataloguing-in-Publication Data
Smyth, A.L.
John Dalton, 1766–1844: A Bibliography of Works by and about him, with an Annotated List of his Surviving Apparatus and Personal Effects. – revised and expanded edition.
1. Dalton, John, 1766–1844—Bibliography. I. Title. II. Manchester Literary and Philosophical Society.
016.5'4092

US Library of Congress Catologing-in-Publication Data
Library of Congress Catolog Card Number is pre-assigned as 97–077127

Set in 11/12pt Garamond
by XL Publishing Services, Tiverton
Printed in Great Britain by Titus Wilson & Son, Kendal

Gifted to Scarborough Meetinghouse
Library by Bromley Davenport.

July 2007.
Religious Society of Friends (Quakers)

JOHN DALTON 〄 1766–1844

Engraving by Worthington from the oil painting of Dalton
by Joseph Allen, 1815 **436**

Dedicated to the memory of

JAMES McCURDY (1924–96)

A generous benefactor of the
Manchester Literary and Philosophical Society

࿊ Contents

❧ List of illustrations

All illustrations not acknowledged above are in the the Manchester Literary and Philosophical Society's collection.

?✦ Introduction

The original edition of this bibliography was published to coincide with the bicentenary of Dalton's birth in 1966 when an international conference was held in Manchester convened by the Royal Society, the Chemical Society, the Society of Chemical Industry and the Manchester Literary and Philosophical Society. Most of the activities were based on the latter Society's newly built premises on the historic site in George Street where Dalton first announced his atomic theory in October 1803 and where he met many of the men whose creative talents made Manchester the first industrial city.

The occasion gave a considerable stimulus to Dalton studies. Under the editorship of Professor D.S.L. Cardwell, the Conference papers were published in 1968. Three major works on Dalton have appeared – by Greenaway (1966), Patterson (1970) and Thackray (1972) – and there have been numerous periodical articles written over the past thirty years. In this time too, there have been many communications from librarians, archivists and Dalton scholars drawing attention to items omitted or newly revealed. One original oil painting has been discovered and a bust of Dalton disinterred. There have also been film and television presentations made about him. Many items have changed location and the necessary updating has been made. Perhaps the most exciting discovery was the ascertaining of Dalton's genetic code by Mollon, Dulai, Hunt and Bowman in 1995, using his preserved eye in the Society's possession, and thus diagnosing the kind of colour blindness from which he suffered.

The aim of this bibliography is to provide in Part One, a complete list of Dalton's published work, his surviving manuscripts, his papers to the Society and his lectures; in Part Two, a list of material about him, including manuscripts, portraits, sculpture, film, television, Daltoniana, separately published items, articles and references in books and periodicals; in Part Three, Dalton's surviving apparatus and surviving personal effects; in Part Four, Dalton's Manchester – a plan of Manchester in 1794 together with a current plan, giving locations relating to Dalton. A new index of correspondence gives locations of related texts; author and subject indexes.

No separate or extensive bibliography of Dalton had been published previously; the following are the main lists of material relating to him:

W. Angus Smith *Memoir of John Dalton* 1856, p. 253–63

J.C. Poggendorff *Biographisch-literarisches Handwörterbuch* 1863, vol. 1, p. 512–14

G. Bugge *Das Buch der grossen Chemiker* 1955, vol. 2, p. 488–9

D.J. Grogan 'John Dalton: a selection of books and papers by and about him' *Manchester Review*, 1957–8, **8**, p. 101–5

J.R. Partington *A History of Chemistry*, 1962, vol. 3, p. 755–6

The Society and John Dalton

John Dalton's close connection with the Manchester Literary and Philosophical Society began when he was introduced to its meetings by Robert Owen and elected a member in 1794. The first of his 117 papers to the Society, he read on 31 October 1794 and the last on 16 April 1844. He was elected Secretary in 1800, Vice-President in 1808 and held the office of President from 1817 to 1844. He used a room in the Society's house as a laboratory in which he did the whole of his experimental work and many of the results of this work he first communicated to members of the Society before publication in the *Memoirs*. The identification of Dalton with the Society continued after his death. His manuscripts and laboratory apparatus came into the possession of the Society, most of his chief biographers have been members of the Society, a Dalton Medal was instituted by the Society and the Council room in the 1960 House was named after him. There is now a considerable body of literature about Dalton and his work and it seems appropriate that the Society should publish a bibliographical survey of what has been written by and about him.

Manuscripts

The great majority of Dalton's manuscripts were presented to the Society in 1864 by William Charles Henry. As well as being Dalton's first biographer, Henry was also his literary executor. As explained in his *Memoirs of John Dalton*, 1854, he was on his way to Italy when Dalton died and the manuscripts were held by Peter Clare, the acting executor. Clare was reluctant to part with them as it is possible that he had some idea of writing a biography himself and, in fact, on one occasion proposed to Henry a joint authorship of Dalton's life. After Clare's decease, the manuscripts were finally forwarded by the surviving executor, William Neild. In the Preface of the *Memoirs*, Henry writes:

> The mss [*sic*] remains of Dalton consist of his early scientific journals, chiefly devoted to meteorology; of note-books, containing records of short excursions in England and Wales; of a correspondence which he had maintained with his brother, Jonathan Dalton…; of a few letters from men eminent in science; and of a letter book, into which he had copied all important letters written by him in the years 1836–42. Of these documents the letters addressed to his brother throw most light on his scientific researches and habits of life.

Henry goes on to mention that he had applied to 'all scientific persons likely to be in correspondence with Dalton but with only limited success.' Several letters from Dalton had been produced by Jonathan Otley and I.F. Crosthwaite (Peter Crosthwaite's grandson) and he had also found letters and other documents among the papers of his father (William Henry).

Further Dalton letters are mentioned by later biographers, Angus Smith, Lonsdale, Roscoe and Brockbank, and many of these are published in their respective works.

In the last decade of the nineteenth century came the outstanding discovery of Dalton's laboratory notes on which Roscoe and Harden's *A new view of the origin of the atomic theory* is mainly based and which contains the following description:

> It may seem remarkable that after a lapse of nearly a century since John Dalton first applied the atomic theory of matter to chemical phenomena, it should be possible to find anything new respecting the genesis of his ideas... The explanation is to be found in the unlooked for discovery, in the rooms of the Literary and Philosophical Society of Manchester, where the whole of Dalton's experimental work was carried out, of his laboratory and lecture notebooks contained in a number of manuscript volumes.
>
> These comprise, in the first place, an extensive series of laboratory notes, commencing in the year 1802, and going down to Dalton's latest years, containing an almost unbroken record of the experimental work to which he so entirely devoted himself... These notes are bound up in twelve volumes, each of these 'compound' volumes being made up of a number of 'simple' notebooks of unruled paper, which have been taken out of their original covers and bound together. They have often been commenced at both ends; some of them have been begun, then left unused for a considerable interval, and finally again brought into requisition. Moreover, several of them seem to have been in use at the same time, appropriated to the experiments on different subjects in progress at the moment. In addition to these very valuable and interesting records, there is also a notebook dated 3rd February 1810 in which are contained the notes of the last six lectures... in the Royal Institution in London.

In 1896, Sir Henry Roscoe presented a cabinet in which to house these volumes.

The next discovery came in 1914 and is described by Professor W.W. Haldane Gee **378**.

> During last summer some members of a special committee of the Council of the Manchester Literary and Philosophical Society, whilst preparing a catalogue of the apparatus and other property in the House of the Society, had their attention drawn to a roll of diagrams in one of the cupboards. Many of these were annotated with the unmistakable handwriting of John Dalton, and there was evidence that they had been made by him and used in his lectures. After further search, the collection reached 150 in number.
>
> In cataloguing the diagrams, it was found that it would be essential to examine the Dalton Manuscripts in the possession of the Society,

especially with the object of finding references to the use of the diagrams. Among the manuscript and miscellaneous packets of papers, a number of Lecture Notes and Syllabuses of Courses of Lectures were found. Hitherto the accounts published of the lectures of Dalton have been very incomplete, and few have realized the important position that his lectures have had in his life-work.

Fortunately, all these finds were well documented, a number of photographs were made and some of the lecture diagrams were copied. Professor Partington in his *History of Chemistry* states that Dr. Harden informed him in 1938 that some of the Dalton manuscripts had then disappeared. These minor losses were followed by catastrophe. More than three-quarters of the collection was destroyed by enemy action on 24 December 1940. The complete destruction of the Society's House was caused by fire extending from adjoining buildings hit by bombs. The historical apparatus, library, furniture and paintings, perished and only the contents of safes in the basement were saved in a more or less charred condition.

The salvaged manuscript items are described in some detail in this bibliography. Many of the charred sheets had to be carefully separated and sorted and their fragile nature made this a lengthy but necessary preliminary to the actual cataloguing of the collection. As the Society was without any premises of its own until 1960, the Council of the Society decided in 1948 to lend some of the surviving material to the Science Museum for display there. This was returned in 1966.

Since the partial destruction of the original collection, the Society has been fortunate in having received various gifts. These include part of the collection formerly at Dalton Hall **323**, a residential hall of Manchester University, to which they had originally been presented by the niece of Robert Benson of Preston. Benson had married Isabella Bewley, who was a grand-daughter of Dalton's cousin, and probably she and her sister Hannah were the nearest relatives at the time of Dalton's death. In 1962, Mr. H.C. Wilson of Kendal presented to the Society a manuscript signed by Dalton, dated 1801, with the title 'New theory of the constitution of mixed aeriform fluids'; this was originally in the possession of Mr. Wilson's uncle, Isaac Braithwaite.

In 1979, the Society faced a difficult decision. Its new building, opened in 1960 to replace the original destroyed in 1940, had major structural problems due to the use of high alumina cement and there appeared to be no easy legal redress. The Society, which is entirely funded by the subscriptions of its members, was going through a period of what seemed intractable financial difficulties and, of necessity, other premises had to be found. Most reluctantly, it was decided to sell the surviving Dalton papers. This was agreed at a special general meeting of the Society with the proviso that every effort should be made to keep them in Manchester. Accordingly, they were offered to the University. Sotheby's gave a valuation of £35,000, although made the comment that at a public auction 'their pre-eminent importance should encourage strong competition and possibly an unexpectedly high price'. The physical condition of the papers, themselves, also reduced the amount of the valuation. The University accepted the papers at the valuation price and money from the National Science Museum, the City of Manchester, the Friends of the National Libraries, and the John Rylands University Library, itself, was used to raise the sum.

At the beginning of the Second World War, manuscripts and other rare items appear to have been stored in metal boxes in the basement of the Society's House. Members, at that time, seemed to have considered the possibility of the collapse of the building caused by bombs but not destruction by the intense heat generated by fires in surrounding textile warehouses resulting from incendiary devices. It was estimated that three-quarters of the Dalton material did not survive (for example, all the items reproduced in Roscoe and Harden (1896) were lost). On the morning of 24 December 1940, the first member to arrive, Mr H. Stevenson, was just in time to prevent the firefighters opening one of the boxes, which would otherwise have had its contents destroyed by spontaneous combustion. Various preservation treatments were subsequently tried on the delicate paper of the surviving manuscripts without much success and, as a precaution, they were microfilmed in 1966.

Thanks to a substantial grant from the North West Region Committee of the Industrial Division of the Royal Society of Chemistry, comprehensive conservation work by the John Rylands Library bindery staff was carried out in 1990–91; this is described in Leitch and Williamson **481.7**. Appropriately, 1991 was the 150th anniversary of the Royal Society of Chemistry and a special exhibition, in which the newly conserved manuscripts were displayed, was held in the Library.

Portraits and Sculpture

Portraits and sculpture have been included in the bibliography although for some items no location has been discovered. References to books and periodicals where there are reproductions have been given. Many derivative portraits and illustrations have been purposely omitted.

Only the most optimistic could have believed that a bust of Dalton would have come to light forty years after the destruction of the old House in which it was displayed. After the 1960–79 building was demolished, the development company which had purchased the land decided to build an office block with much deeper foundations. In the course of preparing the site, a mechanical digger unearthed the Levick bronze **469** presented by Sir Henry Roscoe. This has now been expertly restored and returned to the Society.

As a result of the perspicacity of the staff of the National Portrait Gallery, the Phillips portrait **456** was recognised at a sale in 1987 and purchased for the Gallery. The Dalton Exhibition in 1966 brought to notice a number of portraits not included in the original edition and these are now listed in the portraits section of the present work. No attempt has been made to include all the many derivative portraits of Dalton made mainly in the second half of the nineteenth century.

It must be recorded that the Theed statue **472** erected by public subscription in 1854 in Piccadilly, one of Manchester's busiest centres, was re-sited in 1966 in the comparatively quiet backwater of a forecourt of what is now the Manchester Metropolitan University.

Apparatus and personal effects

Until 1979, most of Dalton's surviving apparatus and other items were displayed in the Dalton Room in the Society's House. With the sale of the building, it was decided that they should be deposited on loan in what is now the Museum of Science and Industry in Manchester. (They were presented outright in 1997). It was thought appropriate to extend the coverage of the Bibliography to include an annotated list of all known existing material of this kind.

Arrangement and form of entries

The form of entry for published works by Dalton **1–14** and separately published works about him **477–492.9** is generally in the style of a full library catalogue entry. Entries for periodical and book references to Dalton's life, 1830–1964 **493–609** and his work 1800–64 **610–771** are arranged chronologically and limited to include only enough bibliographical information to reasonably identify each work. For articles in periodicals and composite works, 1965–95 **800–928**, arrangement is alphabetical by name of author and again the bibliographical information is limited. The section following, 'Dalton in perspective' **929–997**, is arranged in nine subject groups representing organizations or ideas which influenced Dalton or which he influenced.

New to this edition is an index of Dalton correspondence covering known surviving letters together with printed texts of letters quoted in full or in part. It might be necessary to refer to all the sources given in order to build up the complete text of a particular letter.

As the original item numbers have been quoted as references in various publications, it has been decided to retain these numbers. Accordingly, new numbers inserted in the original sequence have been given decimal numbers. It must be emphasised that these are purely a listing device and do not imply the division of a subject. Apparent gaps in the numbering after the decimal point have been made to leave space for future insertions.

Periodical references broadly follow BS5605:1978 and give the title of the periodical (usually in the form of the conventional abbreviation) followed by the year, volume number (in bold), part number (in parenthesis) and inclusive page numbers.

Sources

The following were the principal works consulted in tracing periodical articles and references:

Royal Society *Catalogue of scientific papers*, 1800–1900
Poole, W.F. *An index to periodical literature*, 1802–1907
Internationale Bibliographie der Zeitschriftenliteratur, 1897–date
Essay and general literature index, 1900–33–date
International catalogue of scientific literature, 1901–14
Readers' guide to periodical literature, 1905–date
Industrial arts index, 1913–57
Isis cumulative bibliography, 1913–65 (1–90), 1966–75 (91–100), 1975–84 (101–10)
Subject index to periodicals, 1915–61
Chemical abstracts, **10**, 1916–date
International index, 1916–date
Biography index, 1946–date
British national bibliography, 1950–date
Applied science and technology index, 1958–date
British technology index, 1962–date
British humanities index, 1962–date

The main sources for books were the catalogues of the British Museum (British Library), Library of Congress, and Manchester Reference Library.

Acknowledgements

The compiler gratefully acknowledges the help and advice of all those consulted in the preparation of both editions of this bibliography and particularly the following:

L.L. Ardern, University of Strathclyde, Joint Honorary Librarian of the Society until 1963, who originally conceived the idea of this bi-centenary bibliography and commenced its compilation; Academy of Sciences, U.S.S.R., for providing a list of Russian references and presenting microfilm copies of books and periodical articles on Dalton to the Society; F. Greenaway, Science Museum; R.G. Griffin, Librarian, The Chemical Society; A.E. Jeffreys, Keele University, for information on the location of some Dalton letters; K. Kalaydjieva, Deputy Director, National Library, Sofia; Hilda Lofthouse,

Librarian, Chetham's Library, Manchester; A. Lothian, Librarian, The Pharmaceutical Society; M.R. Parkinson, City Art Gallery, Manchester; C.A. Sizer, Director, City Art Gallery, Salford; R.A. Storey, Assistant Registrar, National Register of Archives; Moses Tyson, Librarian Emeritus, Manchester University.

A great deal of useful material was received in 1965–6 after the original edition had gone to press. Relevant items have been incorporated into this present volume and I am very much obliged to the following who were then in the positions shown –

D.F. James, County Library, Kendal; B.C. Jones, Archivist, Record Office, Carlisle; Sheila J. MacPherson, Archivist in Charge, Record Office, Kendal; Kenneth Smith, City Librarian, Carlisle; Oliver Stallybrass, Librarian, The Royal Institution. I am also very conscious of the great debt I owe to the respective authors of the four books on Dalton mentioned above – Frank Greenaway, formerly of the Science Museum; the late Elizabeth Patterson; Donald Cardwell, Professor Emeritus of the History of Science, UMIST; Professor Arnold Thackray, University of Pennsylvania.

More currently, I would like to record my appreciation for the help given to me by Chris Dickinson, Department of Ophthalmic Optics, UMIST; Peter McNiven, John Rylands University Library, Manchester; Chris Nash, Friends Meeting House, Kendal; John Percy, Salford University Library; G.H. Roberts, Director of Finance, University of Manchester; Jenny Wetton, Museum of Science and Industry in Manchester; the staffs of the Local Studies Unit, the Social Sciences Library and the Technical Library, Central Library, Manchester.

Finally, my grateful thanks go to Janet Allan for the design and production of this book, as well as for much useful advice, and to the Council of the Society for their continued support and encouragement. I am especially indebted to Norman Leece, the Chairman of the Society's Publications Company, for all the time and effort he has enthusiastically given to the arrangements for the publication of this Bibliography and to Stella Lowe, the Honorary Secretary of the Company, for undertaking the onerous task of editing and proof-reading the text.

‿❧ Abbreviations

Periodical titles

Amer. Hist. Rev.	*American Historical Review.* New York
Amer. J. Phys.	*American Journal of Physics.* Lancaster, Pa.
Ann. Chem. Pharm.	*Annalen der Chemie und Pharmacie.* Lemgo and Heidelberg
Ann. Chim.	*Annales de Chimie.* Paris
Ann. Gen. Sci. Phys.	*Annales générales des Sciences Physiques.* Brussels
Ann. Phil.	*Annals of Philosophy.* London
Ann. Phys.	*Annalen der Physik.* Halle and Leipzig
Ann. Sci.	*Annals of Science.* London
Arch. Int. Hist. Sci.	*Archives Internationales d'Histoire des Sciences.* Paris
Arch. Pharm., Ber.	*Archiv der Pharmazie.* Berlin
Bol. Soc. Port. Chim.	*Boletim de Sociedade Portuguesa Chimica.*
Bol. Soc. Quim. Peru	*Bolletin de la Sociedad Quimica del Peru.* Lima
Brit. J. Hist. Sci.	*British Journal for the History of Science*
Brit. J. Ophthal.	*British Journal of Ophthalmology.* London
Brit. J. Radiology	*British Journal of Radiology.* London
Brit. Med. J.	*British Medical Journal.* London
Bull. Instn. Metallurg.	*Bulletin of the Institution of Metallurgists.* London
Chem. & Drugg.	*Chemist and Druggist.* London
Chem. & Ind.	*Chemistry and Industry.* London
Chem. Brit.	*Chemistry in Britain.* Cambridge
Chem. News	*Chemical News and Journal of Physical Science.* London
Chem. 13 News	*Chem. 13 News.* Waterloo, Ontario
Chem. Weekbl.	*Chemisch Weekblad.* Amsterdam
Edinb. J. Sci.	*Edinburgh Journal of Science.* Edinburgh
Edinb. New Phil. J.	*Edinburgh New Philosophical Journal.* Edinburgh
Hist. Sci.	*Historia Scientiarum.* Tokyo
Hist. Stud. Phys. Sci.	*Historical Studies in the Physical Sciences.* Baltimore, Md.
Industr. Engn. Chem.	*Industrial and Engineering Chemistry.* Easton, Pa.
Inzh.-fiz Zh.	*Inzhenerno-fizicheskii Zhurnal.* Minsk
Izv. Akad-Nauk SSSR.	*Izvestiya Akademii Nauk SSSR.* Leningrad
J. Chem. Ed.	*Journal of Chemical Education.* Easton, Pa.
J. Chem. Phys.	*Journal für Chemie und Physik.* Nürnberg
J. Chem. Soc.	*Journal of the Chemical Society.* London
J. Farmac.	*Journal dos Farmaceuticos.* Lisbon
J. Hist. Med.	*Journal of the History of Medicine and Allied Sciences.* New York

J. Inst. Chem.	*Journal and Proceedings of the Royal Institute of Chemistry*. London
J. Mines.	*Journal des Mines*. Paris
J. Nat. Phil.	*Journal of Natural Philosophy, Chemistry and the Arts (Nicholson's Journal)*. London
J. Phys.	*Journal de Physique, de Chimie et de l'Histoire Naturelle*. Paris
J. Prak. Chem.	*Journal für Praktische Chemie*. Leipzig
J. Soc. Biblphy. Nat. Hist.	*Journal of the Society for the Bibliography of Natural History*.
J. Soc. Dy. Col.	*Journal of the Society of Dyers and Colourists*. UK
Khimiya Shk.	*Khimiya v Shkole*. Moscow
Manchr. Rev.	*Manchester Review*. Manchester
Memoirs.	*Memoirs and Proceedings of the Manchester Literary and Philosophical Society*. Manchester
	vols. 1–5. *Memoirs*. 1st series. vols. 1–5 (1785–1802)
	vols. 6–20. *Memoirs*. 2nd series. vols 1–15 (1805–60)
	vols. 21–30. *Memoirs*. 3rd series. vols. 1–10 (1862–87)
	vols. 31–40. *Memoirs and Proceedings*. 4th series. vols 1–10 (1888–96)
	vols. 41–120. *Memoirs and Proceedings*. (1897–1980)
	vols. 121–date *Manchester Memoirs* (1981–date)
	The through numbering of the volumes, adopted in 1897, is the system of reference used throughout this bibliography
Minerva Med.	*Minerva Medica*. Turin
Mitt. Gesch. Med. Naturw.	*Mitteilungen zur Geschichte der Medizin und der Naturwissenschaften*. Hamburg and Leipzig.
Monit. Sc.	*Moniteur Scientifique*. Paris
N.J. Pharm.	*Neues Journal der Pharmacie*. Leipzig
Not. Nat. Heilk.	*Notizen aus dem Gebiete der Natur-und Heilkunde*. Erfurt and Weimar
Notes & Rec. R.S.	*Notes and Records Royal Society*. London
Ohio J. Sci.	*Ohio Journal of Science*. Columbus
Pharm. J.	*Pharmaceutical Journal*. London
Phil. Mag.	*Philosophical Magazine*. London
Phil. Trans.	*Philosophical Transactions of the Royal Society*. London
Pop. Sci. Mon.	*Popular Science Monthly*. New York
Proc. Instn. Mech. Engrs. Lond.	*Proceedings of the Institution of Mechanical Engineers*. London
Proc. R. Inst.	*Proceedings of the Royal Institution*. London

Proc. R. Phil. Soc. Glasg.	*Proceedings of the Royal Philosophical Society of Glasgow.* Glasgow
Proc. Roy. Soc.	*Proceedings of the Royal Society.* London
Proceedings.	*Memoirs and Proceedings of the Manchester Literary and Philosophical Society.* Manchester
	vols. 1–26. *Proceedings.* (1857–60 – 1886–7)
	vols. 31–40. *Memoirs and Proceedings.* 4th series. vols 1–10 (1888–96)
	vols. 41–120. *Memoirs and Proceedings.* (1897–1980)
	vols. 121–date *Manchester Memoirs.* (1981–date)
	In the *Memoirs and Proceedings*, the *Proceedings* are differentiated by the use of Roman numerals in the pagination.
Quart. J.R. Met. Soc.	*Quarterly Journal of the Royal Meteorological Society.* London
Rep. Brit. Assoc.	*Report of the British Association for the Advancement of Science.* London
Roy. Inst. J.	*Journal of the Royal Institution of Great Britain.* London
School Sci. Rev.	*School Science Review.* Hatfield, Herts.
Sci. Amer.	*Scientific American.* New York
Sci. Progr.	*Science Progress.* London
Soc. Philom. Bul.	*Bulletin des Sciences de la Société Philomathique de Paris.* Paris
Stud. Romant.	*Studies in Romanticism.* Boston, Mass.
Trans. Rochdale Lit. Sci. Soc.	*Transactions of the Rochdale Literary and Scientific Society.* Rochdale
Vop. Ist. Est. Tech.	*Voprosy Istorii Estestvoznanya Itekniki.* Moscow
Z. Phys.	*Zeitschrift für Physik.* Brunswick
Z. Phys. Chem.	*Zeitschrift für Physikalische Chemie, Stöchiometrie und Verwandschaftslehre.* Leipzig
Zh. Fiz. Khim.	*Zhurnal Fizicheskoi Khimii.* Moscow.

Locations

APS	American Philosophical Society, Philadelphia
BL	British Library, London
CPL	Carlisle Public Library
DU	Durham University, Prior's Kitchen
EFS	University of Pennsylvania, E.F. Smith Collection
HSP	Historical Society of Pennsylvania, Philadelphia
KFM	Fitz Park Museum, Keswick
KPL	Kendal Public Library

LC	Library of Congress, Washington
LP	Manchester Literary and Philosophical Society
LU	Liverpool University
MAG	City Art Gallery, Manchester
MC	Chetham's Library, Manchester
MGS	Manchester Grammar School
MPL	Central Library, Manchester
MU	Manchester University, John Rylands University Library
NLS	National Library of Scotland, Edinburgh
NPG	National Portrait Gallery, London
RI	Royal Institution, London
RS	Royal Society, London
SF	Society of Friends, London
SM	Science Museum, London
SNPG	Scottish National Portrait Gallery
SU	Salford University
TCC	Trinity College, Cambridge
UStA	University of St Andrews
WHM	Wellcome Historical Medical Library, London
WPL	Warrington Public Library

Other abbreviations

A.L.S.	autograph letter signed
D.N.B.	*Dictionary of National Biography*
diagr.	diagram
facsim.	facsimile
front.	frontispiece
illus.	illustration
mf.	microfilm
ms, mss.	manuscript(s)
port.	portrait
splt.	supplement

Throughout the text, 'the Society' denotes the Manchester Literary and Philosophical Society

Part One ❧ WORKS BY DALTON

❧ Separately published works

1 *Meteorological observations and essays* By **John Dalton**, professor of
mathematics and natural philosophy, at the New College, Manchester.
London, printed for W. Richardson, J. Phillips: Kendal, W. Pennington, 1793.
xvi 208p. diagrs. 20 x 12 cms.
Bound with *Philosophical essays from vol., 5, pt. 2 of the Memoirs* in copies in LC
and MPL
Copy in MPL inscribed 'Jonathan Dalton from the author'

2 *Meteorological observations and essays* By **John Dalton**, professor of
mathematics and natural philosophy at the New College, Manchester.
London, printed for T. Ostell, [n.d.] xvi 208p. diagrs. 22 x 13 cms.
The price on the title page is five shillings compared with four shillings in
the previous entry
MPL

3 *Meteorological observations and essays* By **John Dalton**, D.C.L., F.R.S., president
of the Literary and Philosophical Society, Manchester,... Second edition.
Manchester, printed by Harrison and Crosfield for Baldwin and Cradock,
London, 1834. xx, 244p. tables, diagrs. 21 x 13 cms.
'The second edition is printed verbatim from the first and I have only added
a few notes at the end under the head of *Appendix to the second edition*; and
some observations on clouds, on thunder, and on meteors particularly the
aurora borealis' – preface
MU has author's signed copies presented to Jonathan Dalton, 1834, Jonathan
Otley, Hannah Tipping, Isabella Bewley and Earl Fitzwilliam

4 *Elements of English grammar: or a new system of grammatical instruction, for the use
of schools and academies* By **John Dalton**, teacher of the mathematics and
natural philosophy, and secretary to the Literary and Philosophical Society,
Manchester. London, printed for W.J. and J. Richardson, by R. and W. Dean
and Co., Manchester, 1801. xvi, 122p. 16 x 10 cms.
MU copy has Peter Clare's signature

5 *Elements of English grammar: or a new system of grammatical instruction, for the use
of schools and academies* By **John Dalton**, teacher of the mathematics and
natural philosophy, and secretary to the Literary and Philosophical Society,
Manchester. Second edition. London, printed for T. Ostell by M. Swinney,
Birmingham, 1803. xx, 122p. 16 x 10 cms.

6 *Philosophical essays from vol v. part ii of the Memoirs of the Literary and Philosophical Society, Manchester* By **John Dalton**. Manchester, printed by R. and W. Dean and Co., 1802

This is the common title-page to the following essays each with its own title-page, except the last, and separate pagination

> *Experiments and observations to determine whether the quantity of rain and dew is equal to the quantity of water carried off by the rivers and raised by evaporation: with an enquiry into the origin of springs* By **John Dalton**. Manchester, printed by R. and W. Dean and Co., [n.d.]. 29p. map. 20 x 12 cms.
> See also **26**
>
> *Experiments and observations on the power of fluids to conduct heat: with reference to Count Rumford's seventh essay on the same subject* By **John Dalton**. Manchester, printed by R. and W. Dean and Co., [n.d.]. 28p. 20 x 12 cms.
> See also **27**
>
> *Experiments and observations on the heat and cold produced by the mechanical condensation and rarefaction of air* By **John Dalton**. Manchester, printed by R. and W. Dean and Co., 1801 14p. 20 x 12 cms.
> See also **28**
>
> *Experimental essays on the constitution of mixed gases: on the force of steam or vapour from water and other liquids in different temperatures, both in a Torricellian vacuum and in air: on evaporation: and on the expansion of elastic fluids by heat, etc.* By **John Dalton**. Manchester, printed by R. and W. Dean and Co., 1802. 70p. 1pl. 20 x 12 cms.
> See also **29**
>
> *Meteorological observations*
> [1802]. 9p. 20 x 12 cms. No title-page.
> See also **30**

These essays are bound with the first edition of *Meteorological observations and essays* in copies in MPL and LC.

At the time of Dalton's membership of the Society, volumes of the *Memoirs* were published somewhat erratically – 1793, 1796, 1798, 1802, 1805, 1813, 1819, 1824, 1831, 1842, 1843. Because of the long intervals between volumes, preprints of papers were often produced, usually with a separate title-page and pagination, in sewn paper wrappers. Preprints of some Dalton papers, later than the above, are in RS and MPL. For a complete list of his periodical contributions, see **25–80**.

7 *Remarks on Mr. Gough's two essays on the doctrine of mixed gases: and on Professor Schmidt's experiments on the expansion of dry and moist air by heat* By **John Dalton**. Manchester, printed by S. Russell, 1805. 14p. 19 x 12 cms. Reprint of *Memoirs*, 1805, **6**, 425–36.
See also **39**
MPL

8 *Syllabus of a course of lectures on experimental philosophy* By **John Dalton**.
Manchester, printed by S. Russell, 1808. 24p. 21 x 13 cms.
The detailed syllabus of a course of fifteen lectures.
MPL.

9 *A new system of chemical philosophy* By **John Dalton**. Manchester, printed by S.
Russell for R. Bickerstaff, London, 1808–27. 2 vols. in 3. 8pl., tables. 22 x 13
cms. **vol. 1, pt. 1.** Manchester, printed by S. Russell for R. Bickerstaff,
London, 1808. vii, 220p: **vol. 1, pt. 2.** Manchester, printed by Russell and
Allen for R. Bickerstaff, London, 1810. viii, p. 221–560: **vol 2, pt. 1.**
Manchester, printed by the executors of S. Russell for George Wilson,
London, 1827. x, 375p.
'The work now submitted to the public was begun to be printed in 1817; and
the 13th and 14th sections, containing the oxides and sulphurets, were
printed off before the end of October of the same year. The printing of the
rest of the work to the appendix was finished in September, 1821. One sheet
of the appendix was printed at the end of 1823; but no addition was
afterwards made till May, 1826; when the printing was resumed, and has been
continued to the present time' – preface to vol. **2**, pt. 1. No more published
in this edition.

10 *A new system of chemical philosophy* Facsimile edition of 1,000 copies
published by William Dawson and Sons, Ltd., London, [1953]

10.2 *A new system of chemical philosophy vol. 1* With new introduction by
Alexander Joseph New York, Citadel Press, [1964] (Science classics library)
Reprint of 1808–10 edition
Other reprints: – New York, Philosophical Press [1964]. London, P. Owen
[1965]

11 *A new system of chemical philosophy* By **John Dalton**, D.C.L., LL.D., F.R.S.,
L.&E., M.R.I.A. Second edition pt. 1. London, John Weale, printed by
Simpson and Gillett, Manchester, 1842. vi, ii, 220p. 4pl., tables. 21 x 13 cms.
'The first edition of this part of the work having been out of print for some
years, the author has been induced… to publish a second edition without
making any material alteration to it' – preface. No more published in this
edition.

12 *On oil and the gases obtained from it by heat* By **John Dalton**, F.R.S. etc. Read 6
Oct., 1820. 19p. 21 x 13 cms.
Lacks title page. Preprint of *Memoirs*
See also **64**

13 *On the phosphates and arseniates* By **John Dalton**, D.C.L., F.R.S., etc.
Manchester, John Harrison, printer, 1840. 21p. 1pl. 21 x 13 cms.
'I sent the account of the phosphates and arseniates to the Royal Society, for

their insertion in the Transactions. They were rejected. Cavendish, Davy, Wollaston and Gilbert are no more' – p. 12

14 *On the (phosphates and arseniates) microcosmic salt, acids, bases and water, and a new and easy method of analysing sugar* By **John Dalton**, D.C.L., F.R.S., etc. Manchester, John Harrison, printer, 1840–2. 14p., 10p. 21 x 13 cms. The words 'phosphates and arseniates' are pasted over This title-page is not used when items **13** and **14** are bound in with *A new system of chemical philosophy*, vol. 2, pt. 1 LP

?◦ Translations and edited texts

---

15 *Ein neues System des chemischen Theiles der Naturwissenschaft, von John Dalton* Aus dem Englischen übersetzt von Friedrich Wolff... Berlin, Hitzig, 1812–13. 2 pts. in 1 vol. pl., tables. 21 cms.

15.2 *A new system of chemical philosophy* [in Japanese] Tokyo, Uchida Rokakuho Publishing Co., 1986. 572p.

16 Ostwald, F.W. (ed.) *Die Grundlagen der Atomtheorie* Abhandlungen von J. Dalton und W.H. Wollaston, 1803–8. Herausgegeben von W. Ostwald. Leipzig, Engelmann, 1889. 30p. pl. 19 cms. (Ostwald's Klassiker der exacten Wissenschaften, No. 3) Includes translation of *On absorption of gases by water and other liquids* and extracts from *A new system of chemical philosophy*

17 Ostwald, F.W. (ed.) *Das Ausdennungsgesetz der Gase* Abhandlungen von Gay-Lussac, Dalton, Dulong und Petit, Rudberg, Magnus, Regnault, 1802–42. Herausgegeben von W. Ostwald. Leipzig, Engelmann, 1894. 211p. table, diagr. 19 cms. (Ostwald's Klassiker der exacten Wissenschaften, No. 44). Includes *On the expansion of elastic fluids by heat*

18 D., L. [Dobbin, Leonard?] (ed.) *Foundations of the atomic theory* Comprising papers and extracts by **John Dalton**, William Hyde Wollaston, M.D., and Thomas Thomson, M.D. (1802–8). Edinburgh, The Alembic Club, 1899. 48p. 19 cms. (Alembic Club reprints, No. 2) Includes *Experimental inquiry into the proportion of the several gases or elastic fluids, constituting the atmosphere. On the absorption of gases. On the constitution of bodies. On chemical synthesis*

18.5 Knight, D.M. *Classical scientific papers: chemistry* London, Mills and Boon, 1968. xxiii, 391p. 'The papers reprinted here have been selected to recapture the argument

over theories of matter which raged for the century following the original publication of **Dalton's** atomic theory.'
Includes extracts from *A new system of chemical philosophy, vol. 1, The signification of the world particle* **45,** *Remarks on the essay of Dr Berzelius* **51.**

19 [Smith, W.J.] (ed.) *Foundations of the molecular theory* Comprising papers and extracts of **John Dalton,** Joseph-Louis Gay-Lussac and Amedeo Avogadro (1808–11). Edinburgh, The Alembic Club, 1899. 51p. 19 cms. (Alembic Club reprints, No. 4).
Includes extracts from *A new system of chemical philosophy.*

20 Randall, W.W. (ed.) *The expansion of gases by heat* Memoirs by **Dalton,** Gay-Lussac, Regnault and Chappius. New York, American Book Company, [1902]. vii, 166p, illus., diagrs. 21 cms
Includes – *On the expansion of gases by heat.* Extracts from *A new system of chemical philosophy.* Biographical sketch of Dalton

21 Kedrov, B.M. (ed.) *Compilation of selected works on atomic theory, 1802–10* [in Russian]. Translated from the English by A.A. Liebermann. Moscow, State Scientific and Technical Publishing House for Chemical Literature, 1940. 242p. front. (port.), diagrs.
Includes supplement by B.M. Kedrov, *John Dalton, father of modern chemistry,* p. 151–242
LP. mf.

21.5 Manchester Literary and Philosophical Society *Manuscripts and some printed items in the library of the Society* Microfilm. The originals are now in the John Rylands University Library
MPL Reference: MF883/6,995

❧ Early mathematical and philosophical writings

These take the form of solutions to mathematical questions and philosophical queries in the *Gentleman's Diary* and the *Ladies' Diary*. In this way Dalton established a reputation as a mathematician resulting in his appointment as tutor in mathematics at Manchester New College on the recommendation of John Gough, himself a frequent contributor to the two publications. (See T.T. Wilkinson An account of the early mathematical and philosophical writings of the late Dr. Dalton, *Memoirs,* **17,** 1855, 1–30.) Some of Dalton's own copies are still preserved

22 *The Ladies' Diary: or Woman's Almanack,… containing new improvements in arts and sciences, and many entertaining particulars designed for the use and diversion of the fair-sex*

1784 (81) **Dalton's** name appears amongst those answering the prize enigma (p. 17) and also questions 803 (p. 33), 807 (p.36), 809 (p. 38) and 813 (p. 41)

1785 (82) Similarly, questions 820 (p. 34) and 824 (p. 37)

1787 (84) Question 850 (p. 34) [answered] algebraically by **Mr. John Dalton**, teacher of the mathematics in Kendal. This is the first publication of a solution by **Dalton** and is probably his earliest contribution in print.

1788 (85) Queries 1 (p. 25) Caoutchouc. 3 (p. 25) Sun shining on the fire. 4 (splt.) Use of ring in marriage. Questions 864 (p. 34), 866 (p. 36), 875 (p. 42) and proposes 884 (p. 47)

1789 (86) Queries 1 (p. 23) Sounds at night. 2 (splt.) Singing of a tea-kettle. 3 (p. 24) Divorce. 4 (splt.) Living toads in rocks. 5 (p. 25) Air temperatures. 6 (splt.) Immortality. Questions 882 (p.35) and 885 (p. 38)

1790 (87) Queries 1 (splt.) Friendship and love. 2 (p. 24) Morning sky. 5 (splt.) Yawning. Questions 897 (p. 35), 903 (p. 39) and 907 (p. 44–5)

1791 (88) Questions 908 (p. 33), 916 (p. 39) and 921 (p. 44)

1792 (89) Queries 1 (p. 24) The pleasure arising from conferring an obligation. 2 (p. 24) Second marriage. 3 (p. 25) Solubility of sugar. Question 925 (p34)

1793 (90) Query 5 (splt.) Moon's apparent disc. Question 951 (p. 43)

1794 (91) Query 1 (p. 24) Cause of mist. Questions 963 (p. 40), 965 (p. 43) and proposes 978 (p. 47)

1795 (92) Question 978 (p. 40–1)

23 *The Gentleman's Diary, or the Mathematical Repository;... containing many useful and entertaining particulars peculiarly adapted to the ingenious gentlemen engaged in the delightful study and practice of mathematics*

1783 (43) **Dalton's** name appears amongst those answering questions 486 (p. 35) and 489 (p. 36)

1788 (48) Proposes question 593 (p. 48)

1790 (50) Questions 606 (p. 38), 614 (p. 42) and proposes 631 (p. 48) and 632 (p. 48)

1791 (51) Question 631 (p. 42–3)

1792 (52) Account of observation on eclipse of the sun (p. 2) Questions 642 and 648 and proposes 663

1793 (53) Question 665 and proposes 681

1794 (54) Question 681 and proposes 699

1795 (55) Question 699

24 *The Scientific Receptacle; containing problems, solutions, queries, enigmas, rebuses, charades, anagrams, etc.*

1795 (2) Proposes question 39 (p. 31)

1795 (3) Question 39 (p. 57)

৵ Periodical and encyclopædia contributions

In each entry, the first reference is to the original publication; some subsequent references are to summaries of the original.

25 Extraordinary facts relative to the vision of colours, with observations
Read 31 Oct., 1794. *Memoirs*, 1798, **5**, 28–45; *Edinb. J. Sci.*, 1831, **5**, 88–98
The first paper Dalton read to the Society and the first detailed description of 'colour-blindness' or 'daltonism'.

26 Experiments and observations to determine whether the quantity of rain and dew is equal to the quantity of water carried off by the rivers and raised by evaporation, with an inquiry into the origin of springs
Read 1 Mar., 1799. *Memoirs*, 1802, **5**, 346–72; *Ann. Phys.*, 1803, **15**, 249–78
'The last point, then much debated, was practically settled by Dalton's conclusion that springs are fed by rain. The same paper contained a further development of his theory of aqueous vapour, with the earliest definition of the *dew-point*' – D.N.B.

27 Experiments and observations on the power of fluids to conduct heat, with reference to Count Rumford's seventh essay on the same subject
Read 12 Apr., 1799. *Memoirs* 1802, **5**, 373–97; *Ann. Chim.*, 1802, **45**, 177–81; *Ann. Phys.*, 1803, **14**, 184–98, 293–6; *Roy. Inst. J.*, 1802, **1**, 268–9
'Combated Count Rumford's view that the circulation of heat in fluids is by convection solely' – D.N.B.

28 Experiments and observations on the heat and cold produced by the mechanical condensation and rarefaction of air
Read 27 June, 1800. *Memoirs*, 1802, **5**, 515–26; *Ann. Chim.*, 1802, **45**, 103–7; *Ann. Phys.*, 1803, **14**, 101–11; *J. Mines*, 1802–3, **13**, 257–69
'Contained the understated but important result that the temperature of air compressed to one-half its volume is raised 50°F' – D.N.B.
Dalton's first paper as Secretary of the Society

29 Experimental essays on the constitution of mixed gases; on the force of steam or vapour from water and other liquids in different temperatures, both in a torricellian vacuum and in air; on evaporation; and on the expansion of gases by heat
Read 2, 16 and 30 Oct., 1801. *Memoirs* 1802, **5**, 535–602; *Ann. Chim.*, 1803, **46**, 250–76; *Ann. Phys.*, 1803, **12**, 310–18, 385–95; 1803, **13**, 438–45; 1803, **15**, 1–24, 121–43; *J. Nat. Phil.*, 1802, **3**, 267–71; 1802, **5**, 241–4; *Phil. Mag.*, 1802, **14**, 169–73; *J. Mines*, 1803, **14**, 33–6; *Soc. Philom. Bull.*, 1803, **3**, 189–91.
Gave him at once a European reputation. Consists of four distinct essays. The first expresses the generalisation that the maximum density of a vapour

in contact with its liquid remains the same whether other gases be present or not and the view that the particles of every kind of elastic fluid are elastic only with regard to those of their own kind. The second essay gave the first table of the varying elasticity of steam and described the dew-point hygrometer. The third essay showed the quantity of water evaporated in a given time to be strictly proportional to the force of aqueous vapour at the same temperature, and to be the same in vacuo. The fourth announced the law (arrived at almost simultaneously by Gay-Lussac) that all elastic fluids expand the same quantity by heat. By these discoveries meteorology was constituted a science – from D.N.B.
See also **90**

30 Meteorological observations made at Manchester
Memoirs, 1802, **5**, 666–74; *Ann. Phys.*, 1803, **15**, 197–205

31 Observations concerning the determination of the zero of heat, the thermometrical gradation, and the law by which dense or non-elastic fluids expand by heat
J. Nat. Phil., 1803, **5**, 34–6

32 Correction of a mistake in Dr. Kirwan's essay on the state of vapour in the atmosphere
J. Nat. Phil., 1803, **6**, 118–20

33 On the supposed chemical affinity of the elements of common air; with remarks on Dr. Thomson's observations of that subject
J. Nat. Phil., 1804, **8**, 145–9; *Phil. Mag.*, 1804, **19**, 79–83

34 Observations on Mr. Gough's strictures on the doctrine of mixed gases
J. Nat. Phil., 1804, **9**, 89–92; *Ann. Phys.*, 1805, **21**, 409–16

35 Observations on Mr. Gough's two letters on mixed gases
J. Nat. Phil., 1804, **9**, 269–75; *Ann. Phys.*, 1805, **21**, 420–36

36 Experimental inquiry into the proportion of the several gases or elastic fluids constituting the atmosphere
Read 12 Nov., 1802. *Memoirs*, 1805, **6**, 244–58; *Ann. Phys.*, 1807, **27**, 369–87; *Phil. Mag.*, 1805, **23**, 349–56
'Dalton's first chemical memoir. It disclosed the insight obtained through study of the combinations of oxygen with nitrous gas, into the law of multiple proportions' – D.N.B.

37 On the tendency of elastic fluids to diffusion through each other
Read 28 Jan., 1803. *Memoirs* 1805, **6**, 259–70; *Ann. Phys.*, 1807, **27**, 388–99; *J. Phys.*, 1807, **65**, 68–75; *Phil. Mag.*, 1806, **24**, 8–14

38 On the absorption of gases by water and other liquids

TABLE

of the relative weights of the ultimate particles of gaseous and other bodies.

Hydrogen	1
Azot	4.2
Carbone	4.3
Ammonia	5.2
Oxygen	5.5
Water	6.5
Phosphorus	7.2
Phosphuretted hydrogen	8.2
Nitrous gas	9.3
Ether	9.6
Gaseous oxide of carbone	9.8
Nitrous oxide	13.7
Sulphur	14.4
Nitric acid	15.2
Sulphuretted hydrogen	15.4
Carbonic acid	15.3
Alcohol	15.1
Sulphureous acid	19.9
Sulphuric acid	25.4
Carburetted hydrogen from stag. water	6.3
Olefiant gas	5.3

The earliest printed table of atomic weights 38

Read 21 Oct., 1803. *Memoirs* 1805, **6**, 271–87; *Ann. Phys.*, 1808, **28**, 397–416; *J. J. Phys.*, 1807, **65**, 57–68; *Phil. Mag.*, 1806, **24**, 15–24

p. 287 – first published table of atomic weights – 'the great foundation stone in chemical science.' 'An enquiry into the relative weights of the ultimate particles of bodies is a subject, so far as I know, entirely new. I have lately been prosecuting the inquiry with remarkable success' – p. 286

39 Remarks on Mr. Gough's two essays on the doctrine of mixed gases, and on Professor Schmidt's experiments on the expansion of dry and moist air by heat
Read 4 Oct., 1805. *Memoirs*, 1805, **6**, 425–36

40 Facts tending to decide the question, at what point of temperature water possesses the greatest density
J. Nat. Phil., 1805, **10**, 93–5; *Ann. Phys.*, 1805, **20**, 392–6

41 Extract of a letter from Mr. J. Dalton: on a remarkable Aurora Borealis
J. Nat. Phil., 1805, **10**, 303

42 Remarks on Count Rumford's experiments relating to the maximum density of water
J. Nat. Phil., 1805, **12**, 28–30; *Ann. Phys.*, 1805, **21**, 458–61

43 Investigation of the temperature at which water is of greatest density, from the experiments of Dr. Hope on the contraction of water by heat at low temperatures
J. Nat. Phil., 1806, **13**, 377–81; 1806, **14**, 128–34

44 On muriatic and oxymuriatic acid, in answer to Justus
J. Nat. Phil., 1811, **28**, 157

45 Inquiries concerning the signification of the word Particle, as used by modern chemical writers, as well as concerning some other terms and phrases
J. Nat. Phil., 1811, **28**, 81–8
Meldrum has suggested t hat this paper, written 19 Dec. 1810, is a reply to Davy's Bakerian lecture, 15 Nov. 1810

46 Observations on Dr. Bostock's review of the atomic principles of chemistry
J. Nat. Phil., 1811, **29**, 143–51

47 Remarks on potassium, sodium, etc., in reply to the communication of Justus
J. Nat. Phil., 1811, **29**, 129–33

48 On respiration and animal heat
Read 21 Mar., 1806. *Memoirs*, 1813, **7**, 15–44

49 On the oxymuriate of lime
Ann. Phil., 1813, **1**, 15–23; *J. Chem. Phys.*, 1814, **10**, 445–62

50 Further observations and experiments on the combinations of oxymuriatic acid with lime
Ann. Phil., 1813, **2**, 6–8; *J. Chem. Phys.*, 1814, **11**, 36–44

51 Remarks on the essay of Dr. Berzelius on the cause of chemical proportions
Ann. Phil., 1814, **3**, 174–80; *J. Chem. Phys.*, 1815, **14**, 462–77

52 Vindication of Mr. Dalton's theory of the absorption of gases by water, against the conclusions of Saussure
Ann. Phil., 1816, **7**, 215–23; *Ann. Chim.*, 1816, **1** 357–73

53 On the chemical compounds of azote and oxygen, and on ammonia
Ann. Phil, 1817, **9**, 186–94; 1817, **10**, 38–47, 83–93; *Ann. Chim.*, 1817, **7**, 36–44, 405–11; *Ann. Phys.*, 1818, **58**, 73–91

54 On phosphuretted hydrogen
Ann. Phil., 1818, **11**, 7–9; *Ann. Chim.*; 1818, **7**, 5–7; *J. Chem. Phys.*, 1818, **24**, 325–6

55 On the combustion of alcohol by the lamp without flame
Ann. Phil., 1818, **12**, 245–6; *Ann. Phys.*, 1819, **61**, 337–48

56 Experiments and observations on phosphoric acid, and on the salts denominated phosphates
Read 22 Jan., 1813. *Memoirs*, 1819, **8**, 1–17

57 Experiments and observations on the combinations of carbonic acid and ammonia
Read 19 Mar., 1813. *Memoirs*, 1819, **8**, 18–32

58 On the *vis viva*
Ann. Phil., 1818, **12**, 444–5

59 'Manometer', 'Meteor', 'Meteorology' and articles on subjects relating to chemistry
In **Rees, Abraham**, *The cyclopedia or universal dictionary of arts, sciences and literature*, 1819

60 Remarks tending to facilitate the analysis of spring and mineral waters
Read 1 Apr., 1814. *Memoirs*, 1819, **8**, 52–63; *Not. Nat. Heilk.*, 1823, **3**, 113–8; *Phil. Mag.*, 1821, **58**, 291–6; *Manchester Iris*, 1822, **1**, (19), 145–6
Explains the principles of volumetric analysis

61 Memoir on sulphuric ether
Read 16 Apr., 1819, *Memoirs*, 1819, **8**, 446–82; *Ann. Chim.*, 1820, **14**, 316–9;

Ann. Gen. Sci. Phys., 1820, **4**, 140–3, 269–70; *J. Chem. Phys.*, 1820, **28**, 363–88; *Ann. Phil.*, 1820, **15**, 117–33; *N. J. Pharm.*, 1821, **5**, (2), 300–3

62 Observations on the barometer, thermometer, and rain, at Manchester, from 1794 to 1818 inclusive
Read 13 Nov., 11 Dec., 1818. *Memoirs*, 1819, **8**, 483–509; *Ann. Phil.*, 1820, **15**, 247–59

63 On the solution of carbonate of lime
Ann. Phil., 1822, NS **3**, 316–7

64 On oil and the gases obtained from it by heat
Read 6 Oct., 1820. *Memoirs*, 1824, **9**, 64–82

65 Observations in meteorology, particularly with regard to the dew-point, or quantity of vapour in the atmosphere; made on the mountains in the north of England
Read 9 Feb., 1821. *Memoirs*, 1824, **9**, 104–24

66 On the saline impregnation of the rain which fell during the late storm, December 5th, 1822
Read 13 Dec., 1822, 21 Mar., 1823. *Memoirs*, 1824, **9**, 324–31, 363–72

67 On the nature and properties of indigo, with directions for the valuation of different samples
Read 14 Nov., 1823. *Memoirs*, 1824, **9**, 427–40; *Phil. Mag.*, 1825, **65**, 122–8

68 On the analysis of atmospheric air by hydrogen
Ann. Phil., 1825, NS **10**, 304–6

69 On the constitution of the atmosphere
Read 12 Jan., 1826. *Phil. Trans.*, 1825 (pt. 2), 174–87

70 On the height of the Aurora Borealis above the surface of the earth; particularly one seen on the 29th March, 1826
Read 17 Apr., 1828. *Phil. Trans.*, 1828, 291–302

71 Observations, chiefly chemical, on the nature of the rock strata in Manchester and its vicinity
Read 28 Dec., 1827. *Memoirs*, 1831, **10**, 148–54

72 Summary of the rain, etc. at Geneva, and at the elevated station of the pass of Great St. Bernard, for a series of years; with observations on the same
Read 17 Oct., 1828. *Memoirs*, 1831, **10**, 233–42; *Edinb. New Phil. J.*, 1833, **15**, 101–5

73 Physiological investigations arising from the mechanical effects of atmospherical pressure on the animal frame

Read 8 Jan., 1830. *Memoirs*, 1831, **10**, 291–302; *Edinb. New Phil. J.*, 1832, **13**, 90–6; *Not. Nat. Heilk.*, 1832, **34**, 337–42; *Ann. Chem. Pharm.*, 1832, **4**, 101–9

74 A series of experiments on the quantity of food taken by a person in health, compared with the quantity of the different secretions during the same period; with chemical remarks on the several articles
Read 5 Mar., 1830. *Memoirs*, 1831, **10**, 303–17; *Edinb. New Phil. J.*, 1833, **14**, 62–9; *Not. Nat. Heilk.*, 1833, **36**, 225–31

75 [Newspaper reports of] lecture on the atomic theory to Manchester Mechanics Institution, 19 Oct., 1835
Manchester Courier, 24 Oct., 1835. *Manchester Herald*, 28 Oct., 1835. *Manchester Times*, 24 Oct., 1835. Mentioned by Meldrum, *Memoirs*, 1910–11, **55** (3), 11. See also **362.4**

76 Observations on certain liquids obtained from caoutchouc by distillation
Phil. Mag., 1836, **9**, 479–83; *J. Prak. Chem.*, 1837, **10**, 121–6

77 On the non-production of carbonic acid by plants growing in the atmosphere
Rep. Brit. Assoc., 1837, **6**, pt. 2, 58

78 Sequel to an essay on the constitution of the atmosphere, published in the *Philosophical Transactions* for 1826; with some account of the sulphurets of lime
Phil. Trans., 1837, 347–64; *Phil. Mag.*, 1838, **12**, 158–68, 397–406

79 Observations on the barometer, thermometer, and rain, at Manchester, from the year 1794 to 1840 inclusive, being a summary of the essays on meteorology
Read at various times from the year 1830 to 1840. *Memoirs*, 1842, **11**, 561–89

80 Observations on the various accounts of the luminous arch, or meteor, accompanying the Aurora Borealis of November 3rd, 1834
Read 26 Dec., 1834. *Memoirs*, 1842, **11**, 617–27

?☛ Scientific manuscripts

Lecture notes

Some of the lecture notes were used at more than one series of lectures, probably with revisions. The numbering of lectures in the notes does not always coincide with those in the syllabuses. For a list of Dalton's lectures, see **307**

81 Acoustics
[1791?–1804?] 24 sheets. 33 x 21 cms. to 13 x 20 cms. Lect. 1: Phonics, acoustics or the doctrine of sound. Lect 15: On sound
Also includes material on the theory of harmony. Charred
MU

82 Astronomy
[1791?–1820?] 8 sheets of 32p. 33 x 21 cms. to 23 x 19 cms. Lects. 7–8: Kendal, 1796. Lects. 14–15: [Manchester 1808?] Lect. ?: Apr. 12, 1820 (not in Gee). Lect. 17: Apr. 17, 1820 (not in Gee). Lect: 1: Apr. 27. On the globes. Lect. 15: [n.d.] Lects. 1–3: [n.d.] On the globes
MU

83 Mechanics
[1791?–1818?] 29 sheets. 26 x 20 cms. Lect. 1: Diagrams of experiments. Lect. 2: Mechanic powers. Lect. 6: Pneumatics. Lect. 7: Hydraulics. Lect. 7: Hydraulics and hydrodynamics. Also '*Lectures on natural philosophy*, Kendal, July 25, 1796. 1: mechanics.'
Probably most are the 1818 lectures in Manchester. Charred
MU

84 Optics
[1791?–1820?] 16 sheets of 54p. 33 x 20 cms. to 15 x 20 cms. Lect. 1: [n.d.] Lects. 1–4: [n.d.] Lect. 2: [n.d.] Lect. 5: Apr. 7, 1820 (not in Gee). Lect. 6: Apr. 10, 1820 (not in Gee). Lects. 11–12: [1808?] Of the eye [n.d.] On light [n.d.] Optics [n.d.] Structure of the human eye [n.d.] Diagrams of lenses [n.d.] On colours [n.d.] Delaval on colours [n.d.] Mr. Dalton's ribbands by candlelight [n.d.] Remarks upon the colours by candlelight and by daylight [n.d.]
MU

85 Royal Institution lectures
1810. 56p. of which 14 blank. 21 x 13 cms. Lects. 15–16: Heat. Lects. 17–20: Chemical elements. Conclusion of the course
Printed in Roscoe and Harden **490**, p. 99–125
MU

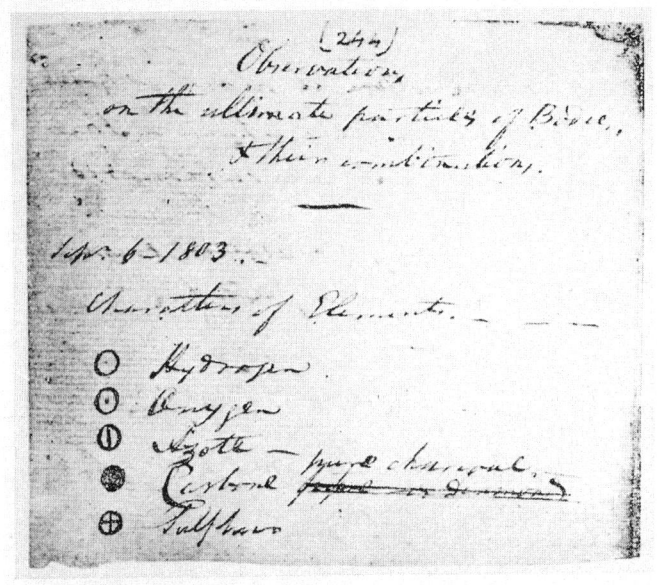

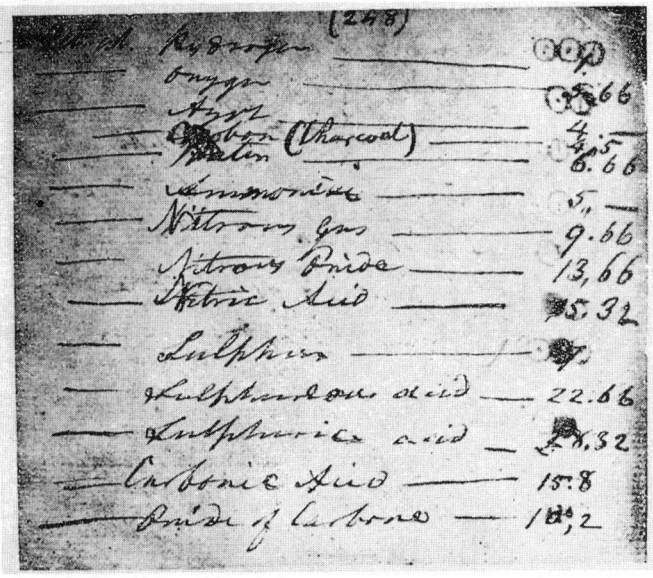

Significant entries in Dalton's notebook, 6 September 1803; the earliest
set of atomic symbols and atomic weights. **490–491.2**

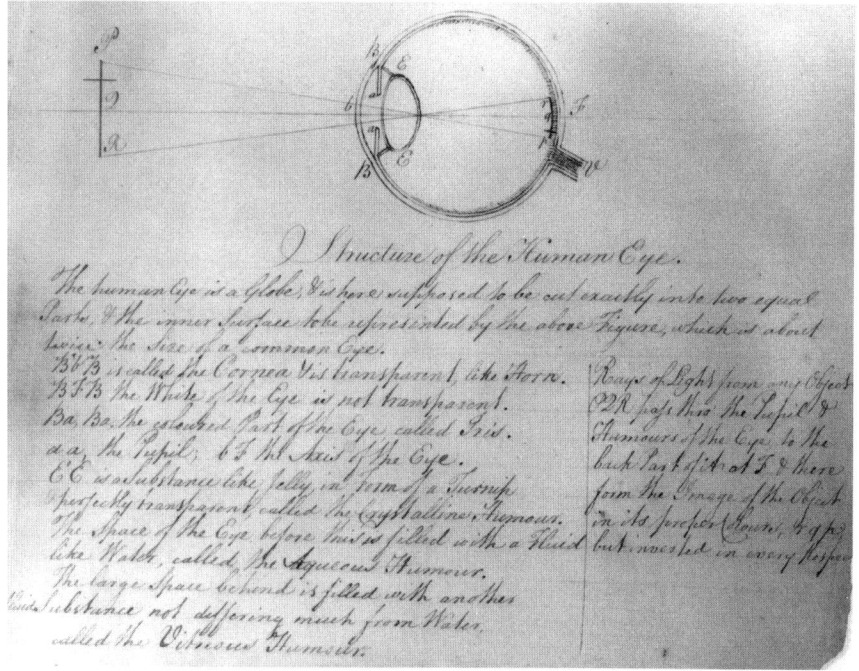

Manuscript of Dalton's lecture on the structure of the human eye, after
conservation work in 1990–1 **84**

86 Natural philosophy, etc.
[Manchester]. Apr. 20, 1814. 1 sheet of 4p. 23 x 9 cms. Introduction. See
Roscoe and Harden **490** p. 125. Charred
MU

87 Pharmaceutical chemistry
[Manchester, 1824–30] 30p. of which 16 blank. 26 x 22 cms. Lect. 9. Dec.,
1830
MU

88 Chemistry
[Manchester, 1827] 46p. 21 x 21 cms. Charred
MU

89 Meteorology
Royal Manchester Institution, 1834. 24p. of which 7 blank. 26 x 18 cms. Lect.
4. Apr. 7, 1834 Lect. 5. Apr. 14, 1834
Some pages are probably part of the 1825 series. Charred
MU

Papers

The numbers of the papers are those given in the list compiled by Angus Smith. See **492**

90 New theory of the constitution of mixed aeriform fluids and particularly of the atmosphere
1801 2 sheets of 6p. 23 x 19 cms. Signed 'John Dalton, Manch[ester], Sept. 14, 1801'. Published in *J. Nat. Phil.*, 1802, **5**, 241–4 [i.e. Oct. 1801] and incorporated in Essay 1 of the '*Experimental essays on the constitution of mixed gases*'. *Memoirs*, 1802, **5**, 538–50. See also **29**
MU

91 Chemical analysis of the mineral waters of Buxton
Read 10 Dec., 1819. Paper 64. 20p. of which 7 blank. 21 x 22 cms. Badly charred
MU

92
Observations on the deutoxide of hydrogen or oxygenized water lately discovered by M. Thenard
Read 1 Nov., 1822. 20p. of which 12 blank. 20 x 22 cms.
Not in Smith's list of papers **492**. Charred
MU

93 On associations for the promotion of the physical sciences, literature and the arts

The Royal Manchester Institution (now the City Art Gallery) where Dalton gave a course of lectures on meteorology in 1834 **89**

Read 15 Oct., 1824. Paper 75. 32p. of which 19 blank. 23 x 22 cms. Charred
MU

94 Results of meteorological observations at Manchester for 31 years with
remarks upon them
Read 15 Apr., 1825. Paper 77. 20p. of which 11 blank. 22 x 22 cms.
The last 3p. describe Dr. Edmund Clarke's observations on Mont Blanc.
Badly charred
MU

95 On the constitution of the atmosphere
Read 30 Dec., 1825. Paper 78. 36p. of which 24 blank. 24 x 21 cms. Also read
at the Royal Society, 24 Feb., 1826. Charred
MU

96 An historical sketch of the Society's library with an account of its present
state
Read 26 Nov., 1827. Paper 81. 32p. of which 14 blank. 24 x 22 cms.
Includes list of librarians and 'expenses attending the library'. Badly charred
MU

97 Observations, chiefly chemical, on the nature of the rock strata in
Manchester and its vicinity
Read 28 Dec., 1827. Paper 82. 16p. of which 8 blank. 24 x 22 cms. Charred
MU

98 Chemical observations on certain atomic weights, as adopted by different
authors, with remarks on the notation of Berzelius
Read 15 Oct., 1830. Paper 87. 16p. of which 1 blank. 22 x 22 cms.
Text in Thackray **492.5** p. 90–108. Charred
MU

99 Salts, oxides, sulphurets
Read 18 Feb., 1831. Paper 91. 92p. of which 14 blank. 26 x 21 cms.
Laboratory notebook. Contains 'On the quantity of oxygen in the
atmospheric air'. Signature inside back cover, Thomas Barton, Blackburn.
MU

100 Sequel to an essay on the constitution of the atmosphere printed in the
Philosophical Transactions, 1826, with some account of the sulphurets of lime
Read 21 Oct. and 4 Nov., 1836. Papers 104 and 105. 34p. of which 9 blank.
25 x 22 cms. Also read with additions before the Royal Society, 15 June, 1837.
Charred
MU

Notebooks, miscellaneous mss.

101 Diophantine problems
[178–] 44p. of which 4 blank. 20 x 17 cms.
MU

102 Meteorological observations
1787–92 and 1803–27. 9 vols. 33 x 21 cms. Register of daily meteorological
observations. Sept. 1787–Dec. 1788; Dec. 1788–Dec. 1789; Jan.–July 1790;
Jan.–June 1791; July–Dec. 1791; Jan.–June 1792; July–Dec. 1792.
MU
1803–1816; 1816–27
SM

103 Mathematical problems and rules of grammar
[1788?] 36p. 20 x 16 cms.
On cover – Notebook No. 1 and illustration with title '*and these are joys of our
dancing days*'. Charred
MU

104 Hortus Siccus, seu Plantarum… in Agris Kendal vicinis sponte
nascentium Specimina, Opere et Studio Joannis Dalton collecta et secundum
Classes et Ordines disposita
1790. ms. 11 pts.
Entry in 'Catalogue of the books of the Manchester Free Library', 1864. See
also **380**

104.1 Hortus Siccus seu plantarum diversarum in agris Kendal vicinis sponte
nascentium specimina. Opere et studio Joannis Dalton collecta
1791–93. ms., 2 vols. This is the herbarium sold to Peter Crosthwaite. Pages
have been added after Dalton's collection and these include several
references to Crosthwaite. An accompanying letter to Prof. H. Balfour of
Edinburgh from Alexander Knight of Keswick, dated 8 October 1870, states
that he bought the volumes at the sale of a local museum. See also **165.3**,
165.5, **165.6** and **926**
Royal Botanic Garden, Edinburgh

105 Genera and species of English plants: those before X found in 1790
36p. of which 2 blank. 10 x 8 cms.
'This index is in the handwriting of the celebrated Dr. Dalton and was
written when he was studying botany' – note inside front cover in Jas.
Crossley's handwriting. Alphabetical index of plants.
MC

105.5 Broughton, Arthur *Enchiridion botanicum* 1782
Contains Dalton's signature and annotations showing months and colours of
plants and flowers
SU

106 A dissertation on the true notion of moving forces and of their use in dynamics: exhibited by an example proposed in the *Comment. Petropolit.* Volume **2d** page 200 by John [Jacques] Bernoulli.
[179–?]. 42p. 25 x 21 cms. Charred
MU

107 Mathematics
[179–] 60p. 25 x 21 cms. Badly charred
MU

108 Meteorological observations and essays
[publ. 1793] 95p. 33 x 20 cms. 30p. 21 x 29 cms.
MU

109 Botanical notes
1798–9. 2 sheets. 33 x 19 cms. Charred
MU

110 Notebook containing notes from Van Nermann's Book, vol. 1785
[electric discharge in water] [5p.] and Geometry [19p.]
[c. 1800] 60p. of which 36 blank. 20 x 13 cms.
MU

111 Globes
[1800?] 64p. of which 40 blank. 20 x 13 cms.
MU

112 Index of notes
1802–13. 40p. of which 24 blank. 18 x 14 cms.
MU

113 Astronomy, use of globes, etc.
[c. 1800–10] 30p. of which 15 blank. 20 x 13 cms.
MU

114 Higgins, William *A comparative view of the phlogistic and anti-phlogistic theories*
2nd ed., 1791
Copy lent to Dalton by Dr. Henry. Contains ms. notes claimed to be in Dalton's hand [1811]. (See Henry's 'Life' p. 78–9)
MU

115 'Manometer', 'Meteor', 'Meteorology' Contributions to **Rees, Abraham,** *The cyclopedia or universal dictionary of arts, sciences and literature,* 1819.
1812. 46 p. of which 19 blank. 25 x 21 cms. Charred
MU

116 A new system of chemical philosophy
[182–]. 8p. 26 x 20 cms.
ms. of p. 347, 349 and 352 of vol. 2, pt. 1, 1827
MU

117 [Notebook] Aurora borealis
10p. of which 7 blank. 28 x 11 cms.
'House expenses' on cover
MU

118 [Notebook] Fundamental points in which we are agreed respecting the collision of elastic balls. Index to No. for Sept. 1830, etc.
28p. of which 10 blank. 21 x 13 cms. Charred
MU

119 [Notebook] Atomic theory and domestic notes 1835–43. 60p. 21 x 16 cms.
MU

120 Notebook on atomic composition of sulphates
60p. of which 53 blank. 29 x 23 cms.
MPL mf.

121 [Laboratory notebook] Salts
20p. of which 10 blank. 28 x 22 cms. Charred
MU

122 Experiments made during a voyage to Bermuda on the carbonic acid on [*sic*] the atmosphere and at Bermuda with sundry meteorological observations at that place [by A. Emmett?]. Communicated by Dr. Dalton [1837?] 20p. of which 10 blank. 27 x 20 cms.
Also contains notes on sulphate of magnesia.
MU

123 [Laboratory notebook]
1839–41. 160p. 26 x 21 cms. Charred
MU

124 On the phosphates
[1840?] 10 sheets. 33 x 20 cms.
Probably in Peter Clare's handwriting. Charred.
MU

125 Arseniates
[1840?] 6 sheets. 33 x 20 cms. Probably in Peter Clare's handwriting. Charred
MU

126 Rain in 1843
1 sheet. 8 cms. x 6 cms.
MU

☙ Non-scientific manuscripts

Accounts

127 Cash accounts
1792–3. 42p. 20 x 16 cms.
Includes lists of instruments and books bought, as well as attendance
register for philosophical lectures, 1791
MU

128 Notebook
1793–5. 48p. 15 x 9 cms.
Includes 'Account of clothing. 1793' and jottings mainly on meteorological
observations
MU

129 Book of accounts
1794 [–1803]. 176p. 19 x 16 cms.
Cover title – Ledger, 1794–1803. Contains alphabetical index of accounts
MU

130 Expenses for journeys
1794–1803. 34p. 15 x 10 cms.
MU

131 Ledger
1803–5. 96p. 25 x 19 cms.
MU

132 Expences [*sic*]
1803, 1806, 1812. 42p. 20 x 12 cms.
Covers house, clothing and pocket expenses
MU

133 Cash accounts
1811–19. 80p. 20 x 17 cms. Charred
MU

134 Expenses
1813–27. 72p. 20 x 12 cms.
Includes names of subscribers to lectures
MU

135 Book of expenses; clothing, pocket expenses
1827. 64p. 20 x 12 cms.
Includes expenses up to 1834 and expenses for various journeys
MU

136 Memoranda, lectures, etc.
1828–34. 52p. of which 16 blank. 20 x 15 cms.
Includes accounts for Manchester lectures, 1828–30
MU

137 Book of expences [*sic*]
1834. 68p. 21 x 14 cms.
Covers period 1834–44
MU

138 Book of accounts: 1. Furniture. 2. Library. 3. Mechanical, chemical and philosophical apparatus. 4. Publications
1835. 86p. of which 36 blank. 20 x 15 cms.
Accounts to 1837 are included. Charred
MU

139 House expenses
1835–7. 100p. 20 x 16 cms.
MU

140 Account book
1837–8. 48p. 26 x 21 cms. 1839–40. 48p. 26 x 21 cms.
MU

Miscellaneous

141 Dilworth, Thomas *The schoolmaster's assistant: being a compendium of arithmetic both practical and theoretical in five parts* 18th ed. 1774. 17 x 10 cms.
Dalton's copy. For a description of the ms. additions to this book, see S.J. Hickson's 'Some early autographs of John Dalton' **381**
MU

142 Law dictionary by J. Dalton
Eaglesfield, Janu. 1st 1781. 22p. 19 x 16 cms.
MU

143 Notes on the prices of various articles of provisions, etc.
Sept., 1791. 14p. of which 11 blank. 16 x 10 cms.
Cover title – Prices of eatables, Kendal, 1785
MU

144 Members of Manchester meeting 3 mo. 1794
1 sheet of 4p. 33 x 20 cms.
Lists names of Quaker families – business, males, females, total, residence.
Dalton is entered as a tutor living in Dawson Street
MU

145 Rules established for the Government of the Literary and Philosophical Society of Manchester
[n.d.] 1 sheet of 4p. 32 x 20 cms.
MU

146 Lines written by Dr. Dalton about the year 1795 under a Profile of Hannah Jepson of Lancaster
1 sheet. 20 x 13 cms. Possibly not in Dalton's handwriting
MU

147 Catalogue of books belonging to the Lit. and Philos. Society of Manchester
1799. 88p. of which 50 blank. 21 x 17 cms.
Includes 'Memorandum of books taken from the library by members of the Society', 1793–1810. Described by Greenaway **479**. Charred
MU

148 School registers and accounts [Jonathan Dalton]
1805–34. 4 vols. 21 x 17 cms.
MU

149 Certificate of attendance at a course of lectures on pharmaceutical chemistry, signed by Dalton
1826
MU

150 Annuities [notes]
[1827?] 18p. of which 5 blank. 30 x 22 cms. Charred
MU

151 Henry Hough Watson. Certificate of attendance at chemistry lectures, 1827–31
1 sheet 22 x 18 cms.
Signed by Dalton, 7 Dec., 1831
MU

152 Receipted account – Bowman
1831. 1 sheet 8 x 19 cms.
14 lessons at 1/6d. £1.1.0.
MU

153 Receipt from Radfords & Co. for the L. & P. Society, 24 Nov., 1837, 2 lamp tops complete
1 sheet 12 x 7 cms.
[signed] John Dalton
MC

154 Place card, British Association meeting
with embossed head of Dalton and signed 'John Dalton, D.C.L., F.R.S., 23rd
of 7th month, 1842.' 12 x 15 cms.
SM

?❧ Letters from Dalton

Over the years, Dalton letters have become widely dispersed to both public and
private collections and the locations of some are not generally known. As an
example, there are five entries below which are based on chance finds in sale
catalogues **156.2, 165.7, 166.2, 179.6, 179.7** and the owners of these items have not
yet been traced. The list is not exhaustive and the Society is always pleased to learn
of any omissions. An index of letters with locations of printed texts is given on
pages 147–154.

155 Abbatt, Robert, Liverpool (second husband of Hannah Tipping)
Manchester, 13 Apr. 1839. 1 sheet of 4p. 25 x 20 cms.
Domestic. Text in Brockbank **477**
MU

156 Abbatt, Robert, Liverpool
Manchester, 14 Sept. 1842. 1 sheet. 23 x 18 cms.
Domestic. Text in Brockbank **477**
SM

156.2 Arago, Dominic, Paris
Manchester, 13 Nov. 1836. 1p., 4to.
Recommends the bearer, William Fairbairn, 'a very large manufacturer of
iron machinery in this place; such as steam engines, boats, boilers, mill
wheels, etc… He is a member of our Philosophical society and has a great
fund of practical mechanical knowledge. He has been very liberal in
furnishing models for one of our members, in order to ascertain the best
form for strength'
Sotheby's catalogue, 5 July 1977, lot 252

157 Babbage, Charles
Manchester, 15 May 1830. 1 sheet. 18 x 11 cms.
Acknowledgement of copy of a publication. Text in Thackray **492.5**
BL

158 Babbage, Charles
Manchester, 7 Dec. 1830. 1 sheet of 2p. 23 x 19 cms..
Election of Herschel to the Royal Society. Text in Thackray **492.5**
BL

159 Backhouse, Jonathan
20 Sep. 1799, 23 Mar. 1801
Two letters about rainfall measurements
DU

159.2 Bernard, T., London
Kendal, 14 July 1804
Bernard was the Royal Institution 'Visitor'. Discusses the possibility of lecturing there the following January. Text in *Memoirs*, 1965–66, **108** 33–34
RI

160 Bewley, George, Whitehaven
Kendal, 9 Apr. 1790
Text in Greenaway **479** and Brockbank **477**
SM

161 Bewley, George, Whitehaven
Kendal, 25 Apr. 1790
Text in Greenaway **479** and Brockbank **477**
SM

161.2 Bewley, George, [Whitehaven] **and Robinson, John**
23 Dec. 1792. 2p.
Dispute with brother over father's will. Listed in Thackray **492.5**.
WHM

161.5 Biot, J.B. [Paris]
5 Aug. 1833
Identical letter is sent to Gay-Lussac. See **170.4**. Listed in Thackray **492.5**
RS Reference: MM18

162 Brockedon, William
Manchester, 20 May 1839. 1 sheet.
Acknowledgement of 'photogenic drawing. It is the best that I have seen'.
See also **440** and **925**
NPG

162.2 Cadell and Davies (publishers), London
Manchester, 1 May 1810
T. Cadell (from 1792 Cadell and Davies) was the publisher of the first five volumes of the Society's *Memoirs* from 1785 to 1802. It seems that there was a dispute over the contract and Dalton, as Vice-President, is writing to say that a new publisher has been appointed. Text in Thackray **492.5**
CPL

162.5 Coxe, J.R. Philadelphia
Manchester, 24 Aug. 1802

Exchange arrangements between the Manchester Society and the American Philosophical Society. See also **346**. Facsim. in Thackray **492.5**
APS

163 Crosthwaite, Peter, Kewsick
Kendal, 21 Feb. 1788. 1 sheet of 4p. 19 x 15 cms.
Refers to exchange of journals and meteorological apparatus. Text in Thackray **492.5**
MU

164 Crosthwaite, Peter, Keswick
Kendal, 21 June 1788. 1 sheet of 4p. 23 x 19 cms.
Meteorological observations. Text in Thackray **492.5**
SM

164.3 Crosthwaite, Peter, Keswick
Kendal, 26 June 1789
Meteorological observations. Text in Thackray **492.5**
WHM

165 Crosthwaite, Peter, Keswick
Kendal, 1 Jan. 1790. 1 sheet. 17 x 20 cms.
On meteorological observations. Text in Thackray **492.5**
LU

165.1 Crosthwaite, Peter, Keswick
Kendal, 20 July 1790
'I am at present pursuing the pleasing study of botany.' Text in Thackray **492.5**
WHM

165.3 Crosthwaite, Peter, Keswick
Kendal, 26 Mar. 1791
Offers to make a collection of dried plants from the Kendal area. Text in Thackray **492.5**
HSP

165.5 Crosthwaite, Peter, Keswick
Kendal, 10 Apr. 1791
'I have at last completed the book of plants.' Text in Thackray **492.5**. See also Wood **926**
WHM

165.6 Crosthwaite, Peter, Keswick
Kendal, 4 Oct. 1791
Completed book of plants. 'I am not so confident of my abilities as to maintain that I have given no plant a wrong name.' Listed in Thackray **492.5**
WHM

165.7 Crosthwaite, Peter, Keswick
Kendal, 23 Dec. 1792
Best way to make certain barometric observations. Refers to Adam Walker, author and inventor, – 'With respect to his character, I never heard any that was a judge, speak of him otherwise than as a boasting pretender. The result of his barometrical observations on Skiddaw, as will be seen from the within, is a strong argument that he was not there.' Also refers to John Gough.
Dawsons Catalogue, 1974, lot 73

166 Crosthwaite, Peter, Keswick
Kendal, 11 Jan. 1793. 1 sheet of 4p. 21 x 17 cms.
Meteorological observations. Text in Thackray **492.5**
SM

166.1 Crosthwaite, Peter, Keswick
Kendal, 18 Apr. 1793
Observations on the aurora. Text in Thackray **492.5**
HSP

166.2 Crosthwaite, Peter, Keswick
Kendal, 22 June 1793
Concerns his *Meteorological observations and essays* and some ideas on refraction in relation to altitude.
Dawsons Catalogue, 1974, lot 73

167 Crosthwaite, Peter, Keswick
Manchester, 2 Aug. 1808. 1 sheet of 4p. 23 x 19 cms
Specimen of tree struck by lightning. Text in Brockbank **477**
MU

167.1 Dalton, Jonathan, Kendal
Manchester, 28 Mar. 1799
Freezing of water. Text in Thackray **492.5**
NLS

167.22 Dalton, Jonathan, Kendal
Devonshire Street, [London], 24 May 1792
Dalton's first visit to London where he attended the Yearly Meeting of the Society of Friends. 'The Town answers my expectation in most Respects but in some vastly exceeds it, particularly in the great Number of Hackney Coaches. …In short this is a most surprising Place… but the most disagreeable Place upon Earth for one of a contemplative turn to reside in constantly.'
LP typescript (see below)

167.24 Dalton, Jonathan, Kendal
Manchester, 6 Mar. 1799

Mainly on meteorological matters. 'I mentioned in one of my letters my preparing a Paper on Springs &c. it was read last sixth day evening (see **26**). …made a thermometer that goes from freezing Mercury to boiling Mercury which I find very useful.' Reports on 'some dreadful accidents by Fire here this Winter. A cotton Factory burned down one evening entirely: damage, 12,000, partly insured.'
LP typescript (see below)

167.26 Dalton, Jonathan, Kendal
Manchester, 6 Feb. 1805
On way to London, stayed a night in Birmingham where he spent some hours with James Watt. Stayed in London three days 'in which time I provided myself the needful apparatus, principally of Jones. …My subscription has gone on with wonderful rapidity [probably to his Manchester lectures]; there are probably nearly 80 on the list, and may be many more before I can be ready'
LP typescript (see below)

167.28 Dalton, Jonathan, Kendal
Manchester, 30 Oct. 1817
Describes an assault in Manchester when his pocket-book was stolen. Reports that his 'highly valued friend, Peter Ewart,' was involved in a very serious accident near Blackburn, as the result of a post chaise overturning. Visited by M. Biot on his way back from the Shetland Isles. Asks if he had mentioned that the French Academy of Sciences had recently elected him a corresponding member. Has received an invitation from the Birmingham Philosophical Society to give a short course of lectures for 80 guineas. 'I have got at least 160 pages of my second volume of chemistry printed.' (see **9**)
LP typescript

The four letters listed immediately above were presented to the Society in 1922 by Colonel Allen of Castle Combe, Chippenham, who believed that they had been purchased at an auction sale. They joined the Society's collection of letters between the brothers dating from 1793 to 1824. All the letters were destroyed in 1940. Henry in his biography **480** gives extracts from three of the letters and there were quotations in an article in the *Manchester City News*, 18 November 1922. This seemed to be all that survived until, in 1995, what appears to be a complete transcription of the four letters was found in an old file, typed on eight sheets of Lit. and Phil. headed paper of the 1920s. The letters are included in *Memoirs*, 1993–4, **132**, 169–76

167.3 Daubeny, C.G.B., [Oxford]
Manchester, 14 Jan. 1841
Acknowledgement of Daubeny's *Supplement*. Listed in Thackray **492.5**
Magdalen College, Oxford, mss 400/88

167.4 Davies, John, Edinburgh
Manchester, 16 January 1823. 1 sheet of 4p. 23 x 19 cms.
Reports a lecture by Milligan on sea-sickness; suggests he conroverts Dr. Wollaston's mechanical principles arising from the motion of the vessel as effecting the human body. Describes the storm on 5 December and the salt water in the rain on which he has written a paper **66**. Includes his account for 56 lectures at 1/6d a lecture
MPL M628/1

167.5 Dickinson, Joseph, Maryport
Manchester, 10 Mar. 1794
Colour blindness
CPL

167.6 Dickinson, Joseph, Maryport
Manchester, 13 Apr. 1794
Colour blindness
CPL

167.7 Dockray, Benjamin
Manchester, 10 June 1822
Plans for a visit to Paris with Dockray. See Henry **480**, 164–8. Text in Thackray **492.5**
MPL

167.8 Emmett, Anthony, Newcastle on Tyne
Manchester, 14 July 1835
Firedamp in mines. Text in Thackray **492.5**
RS Reference: MM 1/9

167.81 Emmett, Anthony, Newcastle on Tyne
Manchester, 15 Dec. 1835
Asks if he would take meteorological observations in Bermuda. Text in Thackray **492.5**
RS Reference: MM 1/10

167.82 Emmett, Anthony, Messrs. Cox and Co., Army Agents, London
Manchester, 13 Feb. 1836
Acknowledges receipt of samples of air and reports from Emmett on aurora of 18 November which was seen in USA on 17th. Urges him to take air samples near the sea and observe any tidal phenomena. Text in Thackray **492.5**
RS Reference: MM1/11

167.9 Ewart, Peter, junior
n.d. Sale of Newton's *Principia*. Listed in Thackray **492.5**
DU

167.94 Faraday, Michael?, London
Manchester, 29 July 1840
Papers on phosphates and arseniates not accepted by *Philosophical Transactions*;
intends to publish them independently. See **13**. Text in Thackray **492.5**
RI

167.95 Faraday, Michael, London
Manchester, 3 Sep. 1840
Effect of lead on health. Text in Thackray **492.5**
RI

168 Fell, John, Ulverston
Manchester, 5 Apr. 1801. 1 sheet of 4p. 26 x 21 cms.
Electrical experiments. Text in Thackray **492.5**
SM

169 Fletcher, John
Manchester, 17 Aug. 1829. 3p.
Account of an ascent of Helvellyn and Saddleback with Jonathan Otley
KFM

170 Fletcher, John, Greysouthen
Cockermouth, 18 July 1824. 2p.
About his portrait. Text in Thackray **492.5**
KFM

170.2 Forbes, J.D., Edinburgh
Edinburgh, 11 Sep. 1834
Proposes to call on Professor Forbes. Text in *Memoirs*, 1973–74, **116**, 10
UStA

170.4 Gay-Lussac, J.L., Paris
Manchester, 5 Aug. 1833
Introduces Major Emmett. Text in Thackray **492.5**
RS Reference: MM 1/7

171 Greenup, Ruth, Greenrig
Kendal, 27 June 1792. 1 sheet of 2p. 21 x 17 cms.
Describes his first visit to London. Ruth Greenup was Dalton's aunt. Text in
Brockbank **477**. See also **167.22**
MU

171.5 Harrison, John, Manchester
Manchester, 11 June 1840
'Please to send in the 3 or 4 sheets that I have sent you. I have altered my
mind. I intend to get copper plates for the symbols and or Lithography.
They are much better and *cheaper*...' Harrison was the local printer

responsible for *Phosphates and Arseniates* **13** and *Microcosmic Salt* **14**
MPL Reference: Brothers collection 9

172 Herschel, Sir John
Manchester, 28 Feb. 1833. Text in Thackray **492.5**
BL

173 Hogg, John, Holbeck
Manchester, 19 Apr. 1825. 1 sheet of 4p. 23 x 19 cms.
Dispatch of copy of *Meteorological essays*
SM

173.2 James, P.M., Birmingham
Manchester, 31 Oct. 1817
Accepts offer of accommodation while in Birmingham. Listed in Thackray
492.5
SF

173.3 James, P.M., Birmingham
Manchester, 14 Apr. 1825
Sending portraits. Concern about boxes not yet received from Birmingham.
Listed in Thackray **492.5**
SF

174 See **Suliot, T.E. 178.2**

174.2 Lord Provost and Magistrates, Edinburgh
Manchester, 14 Dec. 1832
Supports J.D. Forbes as candidate for the Chair of Natural Philosophy,
Edinburgh. Text in *Memoirs*, 1973–74, **116**, 11–2
UStA

174.4 Marshall, John, Hallstead, Ullswater
Manchester, 27 Sept. 1835
Introduces Mrs Sarah Lee who 'intends to study the ichthyology of the Lake
District'. She was the author of *Memoirs of Baron Cuvier* and *The fresh water
fishes of Great Britain*. Text in Thackray **492.5**
TCC Reference: Add mss. c65f36

174.5 Miller, William, Edinburgh
Manchester, 20 Nov. 1827
Approval of proposals for the advertising and distribution of his *New system
of chemical philosophy*. Text in Thackray **492.5**
NLS

174.8 Nicholson, Cornelius, Kendal
Manchester, 10 June 1830
Acknowledges election to Kendal Natural History and Scientific Society.

Listed in Thackray **492.5**
KPL

175 Orchard, [Thomas, attorney?]
London, 28 Dec. 1814
Paying off an annuity
BL

175.5 Robinson, Elihu, Eaglesfield
Manchester, 27 Jan. 1798
Reports on tour to Shrewsbury, Coalbrookdale, Stourbridge, Bridgnorth, Hagley, Birmingham and Oxford. Text in Thackray **492.5**
SF

176 Robinson, John
— 1792. 1 sheet of 2p. 21 x 17 cms.
MU

176.3 Savage, N. London
Manchester, 24 Nov. 1803
Savage was Clerk of the Royal Institution and Dalton asks about arrangements for his 1803 lectures at the Institution. Text in *Memoirs*, 1965–66, **108**, 31
RI

176.5 Sharpe, John, London
Manchester, 20 Nov. 1824
Acknowledges Sharpe's letter of 15 November 1824. **343** Referring to the Lit. and Phil., Dalton writes 'In regard to number… I think we are 130 or more members; but as usual the labouring ones are not very numerous'. Sharpe's paper on steam (*Memoirs* **7**, 1–14) is mentioned. *A new system of chemical philosophy* is dedicated to him. Text in Thackray **492.5**
WHM

177 Sibson, Edmund, Warrington
Manchester, 29 Mar. 1814
Loan of book to Sibson and possibility of him giving a paper to the Society.
WPL

178 Sibson, Edmund, Warrington
Manchester, 7 July 1825
Domestic
WPL

178.2 Suliot, T.E., Leeds
Manchester, 17 Mar. 1837
Possible dates for an interview. Listed in Thackray **492.5**
SM

179 Taylor, Mary, (Moston, Manchester) Geneva
Manchester, 13 June 1825
Discusses visit to Paris. Mentions recent visits to Birmingham and London
where he gave evidence before a committee of the House of Commons.
Would like to have a sample of air from mount [sic] Blanc. Text in *Memoirs*,
1989–90, **129**, 141–3. This differs slightly from that in Brockbank **477** which
is taken from Dalton's copy letter.
MC Copy letter in MU

179.2 Taylor, Mary, Moston, Manchester
Manchester, 2 May 1829 (or 1835)
Dr Wollaston's method of avoiding seasickness. Text in *Memoirs*, 1989–90,
129, 143
MC

179.4 Taylor, Mary, Moston, Manchester
Manchester, n.d.
Hints on a tour of the Lake District. Text in *Memoirs*, 1989–90, **129**, 145
MC

179.5 Thomson, Thomas?, Edinburgh?
Manchester, 2 Mar. 1807
Proposes to spend two or three weeks in Edinburgh and deliver a short
course of four or five lectures. Text in Thackray **492.5**
WHM

179.6 Thomson, Thomas, Glasgow
Manchester, Oct. 1822
Introduces his pupil, John Davies, about to spend the ensuing season in
Edinburgh and Glasgow
Myers catalogue 1977 no. 10 Lot 311

179.7 Thomson, Thomas, Glasgow
Manchester, July 1830
Acknowledges Thomson's new volume on heat and electricity. Gives figures
concerning experiments in theory of atmosphere
Myers catalogue 1977 no. 10 Lot 311

179.9 Thomson, Thomas [Liverpool Quaker]
Manchester, 28 Mar 1834
Sale of collection of minerals. Listed in Thackray **492.5**
SF

180 Tipping, Hannah, Liverpool
Manchester, 8 Apr. 1835. 1 sheet of 4p. 25 x 20 cms.
Domestic. Hannah Tipping was a granddaughter of Dalton's cousin, George
Bewley. She first married Robert Tipping, a tea merchant, and her second

marriage was to Robert Abbatt. Text in Brockbank **477**
MU

181 Tipping, Hannah, Liverpool
Manchester, 31 May 1835. 1 sheet of 4p. 25 x 20 cms.
Domestic. Text in Brockbank. **477**
MU

181.2 Tipping, Hannah, Liverpool
Manchester, 18 Apr. 1836
Offers accommodation in Faulkner Street; describes journey home from
British Association meeting in Dublin. Text in Thackray **492.5**
EFS

182 Tipping, Hannah, Liverpool
Manchester, 12 Sept. 1836. 1 sheet of 4p. 23 x 18 cms.
Domestic. Text in Brockbank **477**
MU

183 Tipping, Hannah, Liverpool
Manchester, 10 Apr. 1838. 1 sheet. 23 x 19 cms.
Domestic
SM

184 Tipping, Hannah, Liverpool
Manchester, 12 Apr. 1842. 1 sheet of 4p. 23 x 18 cms.
Domestic
MU

185 Watson, H.H.
Manchester, 29 July 1834. 1 sheet. 21 x 14 cms.
Domestic
MU

185.4 Wilkinson, Thomas
Kendal, 27 Aug. 1794
Observations of heights of various places in Lake District. Text in Thackray
492.5
WHML

186 Williamson, James, Leeds
Manchester, 8 Nov. 1823. 1 sheet of 4p. 23 x 19 cms. Text in Thackray **492.5**
SM

186.2 Williamson, James, Leeds
Manchester, 12 Apr. 1825
Declines request to lend apparatus. Text in Thackray **492.5**
EFS

187 Wood, – (Miss)
Cockermouth, 14 Sep. 1838. 1 sheet of 4p. 23 x 19 cms.
Domestic
MU

188 Analysis of spring water
Manchester, 14 May 1839. 1 sheet. 25 x 20 cms.
Charge 10 shillings
Correspondent not given.
MU

189 Not being able to get a place in one of the early coaches...
1 sheet. 10 x 11 cms.
A.L.S. undated without name of correspondent.
MPL

189.5 I have been thinking it would be more agreeable to you... if you were
present at the reading of your paper...
A.L.S. undated without name of correspondent
MPL CD

?❧ Papers read before the members of the Society

The following list is based on the entries appearing in **R. Angus Smith's**
'Memoir of John Dalton', p. 253–61. These were extracted from the *Journals* of
the Society. (The *Journals* consisted of twelve manuscript volumes, covering
the period 1781 to 1864; they were destroyed in 1940.) It will be noticed that
the dates quoted here and those given with the corresponding papers in the
Memoirs do not always coincide although the discrepancy is never more than
a month or so. This may be because of the rules introduced in January 1851,
four years prior to the publication of Smith's *Memoir*, which were made so
that the 'scientific world will be assured of the precise date of each paper
and that proper credit will be given to the author in cases of disputed
discovery'. With this sentiment in mind, Smith might have preferred to give
the date of official receipt of a paper rather than that when it was actually
read.

Angus Smith makes the point that all the papers were not viewed by
Dalton as important and some were merely given in all probability to supply
an occasional want of material at the meetings of the Society.

190 Extraordinary facts relating to the vision of colors, with observations
Read 31 Oct. 1794. Printed **25**

191 On the color of the sky, and the relation between solar light and that

derived from combustion; with observations on Mr. Delaval's theory of colors
Read 27 Nov. 1795

192 Essay on the mind, its ideas, and affections; with an application of principles to explain the economy of language
Read 7 Apr. 1798

193 A paper, containing experiments and observations, to determine whether the quantity of rain and dew is equal to the quantity of water carried off by the rivers and raised by evaporation; with an inquiry into the origin of springs
Read 1 Mar. 1799. Printed **26**

194 Experiments and observations on the power which fluids possess of conducting heat; with reference to Count Rumford's seventh essay
Read 12 Apr. 1799. Printed **27**

195 On the color of the sky, and the relation betwixt solar light and that derived from combustion; with observations on Mr. Delaval's theory
Read 7 June 1799

196 Experimental essays, to determine the expansion of gases by heat, and the maximum of steam or aqueous vapour, which any gas of a given temperature can admit of; with observations on the common and improved Steam Engines
Read 18 Apr. 1800

197 On the heat and cold produced by the mechanical condensation and rarefaction of air
Read 27 June 1800. Printed **28**

198 Philological inquiry into the use and signification of the auxiliary verbs and participles of the English language
Read 17 Oct. 1800

199 Review of Dr. Herschel's experiments on the radiant heat, and the reflectibility and refrangibility of light
Read 12 Dec. 1800

200 Read part 1st of Mr. Dalton's paper on the constitution of mixed gases, etc.
Read 31 July, 1801. Printed **29**

201 Read part 2nd of Mr. Dalton's paper on the force of steam, etc.
Read 2 Oct. 1801. Printed **29**

202 Read part 3rd of Mr. Dalton's paper on evaporation, etc.
Read 16 Oct. 1801. Printed **29**

203 On the general causes, force, and velocity of winds; with remarks on the seasons most liable to high winds
Read 22 Jan. 1802

204 On the proportion of the several gases or elastic fluids, constituting the atmosphere; with an inquiry into the circumstances which distinguish the chymical and mechanical absorption of gases by liquids
Read 29 Oct. 1802. Printed **36**

205 On the spontaneous intercourse of different elastic fluids, in confined circumstances
Read 14 Jan. 1803. Printed **37**

206 On the absorption of gases by water
Read 7 Oct. 1803. Printed **38**

207 On the law of expansion of elastic fluids, liquids, and vapours
Read 4 Nov. 1803

208 A review and illustration of some principles in Mr. Dalton's course of lectures on natural philosophy, at the Royal Institution, in January, 1804
Read 24 Feb. 1804

209 On the elements of chemical philosophy
Read 3 Aug. 1804

210 On heat
Read 5 Oct. 1804

211 Review of **Dr. Hope's** paper On the contraction of water by heat
Read 30 Nov. 1804

212 Remarks on **Mr. Gough's** two essays on mixed gases, and on **Mr. Schmidt's**, On moist air
Read 2 Sep. 1805. Printed **39**

213 On respiration and animal heat
Read 7 Mar. 1806. Printed **48**

214 On the constitution and properties of sulphuric acid
Read 6 Feb. 1807

215 On heat
Read 2 Oct. 1807

216 On the expansion of bodies by heat
Read 16 Oct. 1807

217 On the specific heat of bodies
Read 22 Jan. 1808

218 On the specific heat of gaseous bodies
Read 18 Mar 1808

219 On the measure of mechanical force
Read 2 Dec. 1808

220 On respiration
Read 16 Dec. 1808

221 On evaporation
Read 10 Mar. 1809

222 On the compounds of sulphur
Read 7 Apr. 1809

223 On muriatic acid
Read 3 Nov. 1809. Probably printed **44**

224 On sulphuric acid
Read 1 Dec 1809

225 On fog
Read 9 Mar 1810

226 Appendix to his remarks on respiration and animal heat
Read 16 Nov. 1810

227 On hygrometry
Read 28 Dec. 1810

228 On meteorology
Read 3 Apr. 1812

229 Meteorology continued
Read 17 Apr. 1812

230 On the oxymuriate of lime
Read 2 Oct. 1812. Printed **49**

231 Experiments on phosphoric acid and the phosphates
Read 8 Jan. 1813. Printed **56**

232 Experiments and observations on the different compounds of carbonic
acid and ammonia
Read 5 Mar. 1813. Printed **57**

233 On the combinations of gold
Read 1 Oct. 1813

234 Continuation of the paper on the combinations of gold
Read 15 Oct. 1813

235 The combinations of platina
Read 12 Nov. 1813

236 On the cause of chemical proportion, being remarks on a paper by Berzelius
Read 10 Dec. 1813. Printed **51**

237 Experiments on certain frigorific mixtures
Read 7 Jan. 1814

238 Remarks tending to facilitate the analysis of spring and mineral waters
Read 18 Mar. 1814. Printed **60**

239 On metallic oxides
Read 7 Oct. 1814

240 On metallic oxides (continued)
Read 2 Dec. 1814

241 Critical remarks on some modern chemical phrases
Read 27 Jan. 1815

242 Remarks on Saussure's essay on the absorption of gases by liquids
Read 17 Nov. 1815. Printed **52**

243 On the chemical compounds of azote and oxygen
Read 4 Oct. 1816. Printed **53**

244 An appendix to the essay on chemical compounds of azote and oxygen
Read 13 Dec. 1816

245 On phosphurets, or the combinations of phosphorus with earths, alkalies, metals, etc.
Read 3 Oct. 1817

246 Observations on oxides and sulphurets
Read 21 Nov. 1817

247 Observations on the quantity of rain during the last twenty-five years; with remarks on the theory of rain
Read 13 Nov. 1818. Printed **62**

248 Summary of observations on the barometer and thermometer, made at Manchester for the last 25 years
Read 11 Dec. 1818. Printed **62**

249 Experiments on the force of the vapour of ether, to show the fallacy of some of **Dr. Ure's** statements just published in the *Philosophical Transactions*
Read 8 Jan. 1819

250 On sulphuric ether
Read 16 Apr. 1819. Printed **61**

251 On alloys, particularly those of copper and zinc, and copper and tin
Read 15 Oct. 1819

252 On amalgams, and other metallic alloys
Read 12 Nov. 1819

253 A chemical analysis of the mineral waters of Buxton
Read 10 Dec. 1819. ms. **91**

254 On oil, and the gases obtained from it by heat
Read 6 Oct. 1820. Printed **64**

255 On alum
Read 1 Dec. 1820

256 On meteorology, or observations on the weather for the years 1819 and 1820, in Manchester
Read 26 Jan. 1821

257 Observations on meteorology, particularly with regard to the dew point, etc., or quantity of vapour in the air
Read 9 Feb. 1821. Printed **65**

258 Some observations on the salts and sulphurets of iron
Read 5 Oct. 1821

259 On the effects of continued electrification on compound and mixed gases
Read 30 Nov. 1821

260 Observations on the deutoxide of hydrogen or oxygenized water lately discovered by M. Thenard
Read 1 Nov. 1822. ms. **92**

261 On the saline impregnations of the rain which fell during the late storm, viz., December 5th, 1822
Read 13 Dec. 1822. Printed **66**

262 Appendix to an essay on salt rain (read 13 December 1822) with additional observations on the succeeding storms of wind and rain
Read 21 Mar. 1823. Printed **66**

263 On the nature of properties of indigo; with directions for the valuation of different samples
Read 14 Nov. 1823. Printed **67**

264 On various alloys of tin, zinc, lead, bismuth, antimony, etc.
Read 26 Dec. 1823

265 On associations for the promotion of the physical sciences, literature, and the arts
Read 15 Oct. 1824. ms **93**

266 An account of some experiments to determine the light and heat given out by the combustion of different gases
Read 12 Nov. 1824

267 Results of meteorological observations at Manchester, for thirty-one years; with remarks upon them
Read 15 Apr. 1825. ms **94**

268 On the constitution of the atmosphere
Read 30 Dec. 1825. Printed **69**. ms **95**

269 On the height of the aurora borealis above the surface of the earth, particularly the one seen on the 29th of March, 1826
Read 6 Oct. 1826. Printed **70**

270 An appendix to a paper read on October 6th, on the height of the aurora borealis above the surface of the earth
Read 4 Nov. 1826. Printed **70**

271 An historical sketch of the Society's library; with an account of its present state
Read 26 Nov. 1827. ms **96**

272 Observations, chiefly chemical, on the nature of the rock strata in Manchester and its vicinity
Read 28 Dec 1827. Printed **71**. ms **97**

273 Summary of the rain, etc., at Geneva and at the elevated station of St. Bernard, for a series of years, from the 'Bibliotheque Universelle' for March, 1828; with observations on the same
Read 17 Oct. 1828. Printed **72**

274 Physiological investigations, deduced from the mechanical effects arising from atmospherical pressure on the animal frame
Read 8 Jan. 1830. Printed **73**

275 Remarks on a statement of the amount of rain fallen at different places on the line of the Rochdale Canal
Read 22 Jan. 1830

276 On the quantity of food taken by a person in health, compared with the quantity of the different secretions during the same period; with chemical

remarks on the several articles
Read 5 Mar. 1830. Printed **74**

277 Chemical observations on certain atomic weights, as adopted by different authors; with some remarks on the notation of Berzelius
Read 15 Oct. 1830. ms **98**

278 Observations on the causes of colouring matter
Read 29 Oct. 1830

279 Chemical observations on certain atomic weights, as adopted by different authors; with remarks on the notation of Berzelius
Read 23 Nov. 1830. ms **98**

280 Meteorological observations for a period of thirty-seven years; with theoretical remarks
Read 21 Jan. 1831. Printed **79**

281 On the quantity of oxygen in atmospheric air
Read 18 Feb. 1831. ms **99**

282 On the proportion of oxygen gas in the atmosphere
Read 2 Dec. 1831

283 A summary of meteorological observations, for 1831, made in Manchester and the vicinity
Read 13 Jan. 1832. Printed **79**

284 Dr. Dalton's remarks on the meteorology of the last year
Read 11 Jan. 1833. Printed **79**

285 Observations on the anomalous vision of colours
Read 8 Mar. 1833

286 A description of an imaginary aurora borealis in the north of England
Read 1 Nov. 1833

287 An account of meteorological observations, at Manchester and other places, in the year 1833
Read 7 Feb. 1834. Printed **79**

288 Some remarks on clouds, their nature, height, etc.
Read 7 Mar. 1834

289 Observations on certain liquids obtained from caoutchouc by distillation
Read 17 Oct. 1834. Printed **76**

290 Observations on the various accounts of the luminous arch or meteor accompanying the aurora borealis of November 3rd, 1834
Read 26 Dec. 1834. Printed **80**

291 Account of meteorological observations, made in Manchester and other places in 1834
Read 20 Feb. 1835. Printed **79**

292 Read a paper by Dr. Dalton (subject not named in the *Journal*)
Read 2 Oct. 1835

293 An account of meteorological observations, made in Manchester and other places in 1835
Read 15 Feb. 1836. Printed **79**

294 Sequel to an essay on the constitution of the atmosphere; read to the Society in the year 1825. Part 1
Read 21 Oct. 1836. Printed **78**. ms. **100**

295 2nd Part of a paper entitled 'Sequel to an essay on the constitution of the atmosphere'
Read 4 Nov. 1836. Printed **78**. ms. **100**

296 On arseniates and phosphates
Read 2 Oct. 1838. Printed **13**

297 Some account of meteorological observations, made in Manchester, in the years 1836–37–38
Read 5 Feb. 1839. Printed **79**

298 On the ammoniaco-magnesian phosphate, as it was formerly called; or the tribasic phosphates of magnesia and ammonia, as Professor Graham has called it. And on the phosphate of soda and ammonia, or microscopic salt, as it was formerly called; and now tribasic phosphate of soda and ammonia and water, of Professor Graham
Read 1 Oct. 1839

299 On the quantity of acids, bases, and water in the different varieties of salts; with a new method of measuring the water of crystallization
Read 31 Mar. 1840. Printed **14**

300 Some account of meteorological observations, made in Manchester, in the year 1839
Read 29 Apr. 1840. Printed **79**

301 Continuation of a paper on the quantity of acids, bases, and water in the different varieties of salts
Read 6 Oct. 1840. Printed **14**

302 Meteorological observations, made in Manchester and the neighbourhood during the year 1840, or previously
Read 12 Jan. 1841. Printed **79**

303 On a new and easy method of analyzing sugar
Read 9 Mar. 1841. Printed **14**

304 On the citric acid, the oxalic acid, the acetic acid, and the tartaric acid
Read 5 Oct. 1841

305 Meteorological observations, at Manchester, made in the year 1842
Read 20 Jan. 1843

306 On the fall of rain, etc., etc., in Manchester, during a period of 50 years
Read 16 Apr. 1844

?☞ Lectures

---

307 Based on W.W. Haldane Gee's 'John Dalton lectures and lecture
illustrations', *Memoirs*, 1914–15, **59** (12). See also 'Lectures' in subject index
1787 Kendal. Natural philosophy
1791 Kendal. 1 & 2 Mechanics. 3 & 4 Optics. 5 & 6 Hydrostatics. 7 Fire. 8, 9
 & 10 Astronomy. 11 & 12 Use of the globes
1796 Kendal. 12 lectures of which 6 on chemistry
1803–04 London, Royal Institution. Mechanics, electricity, magnetism,
 optics, astronomy, use of globes, sound, heat, constitution of mixed
 gases, meteorology
1805 Manchester. Matter, motion and mechanical principles (2), hydrostatics
 (1), pneumatics (2), hydraulic and pneumatic instruments (1), electricity
 and galvanism (3), magnetism (1), optics (2), heat (2), elements of
 bodies and their composition (1), mixed elastic fluids and the
 atmosphere (1), the absorption of gases by water, etc. (1), meteorology
 (1), astronomy (2)
1806 Manchester. 1805 course repeated
1807 Edinburgh, Glasgow. Heat (2), chemical elements (3)
1808 Manchester. 1 & 2 Matter, motion and mechanics. 3 Hydrostatics. 4 &
 5 Pneumatics. 6 Hydraulics. 7 Steam engine. 8 & 9 Electricity. 10
 Galvanism. 11 & 12 Optics. 13 Meteorology. 14 & 15 Astronomy.
1809–10 London, Royal Institution. 1 Introduction. 2 & 3 Laws of
 motion. 4 & 5 Pneumatics. 6 Hydrostatics. 7 & 8 Steam engine. 9 & 10
 Electricity. 11 & 12 Meteorology. 13 & 14 Astronomy. 15 & 16 Heat.
 17–20 Chemical elements
1811 Manchester, the Lecture Room of the Literary and Philosophical
 Society. 20 lectures on mechanics, hydrostatics, pneumatics, hydraulics,
 electricity, galvanism, optics, meteorology, astronomy and chemistry
 (including heat)
1814 Manchester. Natural philosophy and chemistry

1817 Birmingham Philosophical Institute. Chemistry
1818 Manchester (also given at the Birmingham Philosophical Institute). Mechanics, etc. (15)
1820 Manchester, the Lecture Room of the Literary and Philosophical Society. 1–3 Electricity. 4 Galvanism
1823 Leeds Philosophical and Literary Society. Mechanics (4), meteorology (2)
1824 Manchester Medical School. Pharmaceutical chemistry (15) (Dalton continued his connection with the Medical School for at least six sessions)
1825 Manchester. (also given at the Birmingham Philosophical Institute). Meteorology (6)
1827 Manchester. heat and chemistry (Repeated 1828 and 1829)
1834 Royal Manchester Institution. Meteorology (6?)
1835 Manchester Mechanics' Institution. Meteorology (5) Atomic theory

Part Two ❧ WORKS ABOUT DALTON

❧ Manuscripts

308 Dalton, Jonathan Meteorological observations at Kendal. 1793–1809.
16 vols. 20 x 16 cms.
Continuation of John Dalton's observations, 1788–93
MU

309 Membership certificates of the Manchester Literary and Philosophical
Society signed by Dalton
John Rothwell, 2 May 1800
Illustration in Thackray **492.5**
Prof Arnold Thackray, Wayne, Penn.
H.H. Birley, 27 Apr. 1804
Illustration – *Memoirs*, 1961–2, **104**, f. p. 65
MPL
G.W. Wood, 24 Apr. 1807
LP
J.R. Beard, 26 Jan., 1827
LP

310 We the circumscribed members of the Literary and Philosophical
Society of Manchester as a mark of respect for our learned and worthy
president request that he will do us the favor to sit for his portrait to Mr.
Allen
Round robin signed by 24 members, 1814? 1 sheet. 23 x 18 cms.
MC

311 John Dalton. Pass book, Manchester Bank, Heywood Bros. & Co.
1822–40. 48p. 16 x 10 cms.
MU

312 Dalton Testimonial Committee Cash books, correspondence, etc.
1833–42
Unsorted. Charred
MU

313 Dalton Testimonial Committee Minute book 1833–42. 75p. 22 x 18
cms.
Includes printed resolutions, etc. dated 8 Aug. 1833, 21 Aug. 1833, 23 Aug.
1833, 2 Sept. 1833, 16 Sept. 1833, 19 Sept. 1833, 28 Sept. 1833, 2 Oct. 1833, 4
Oct. 1833, 5 Oct. 1833, 13 Jan. 1834, 21 Feb. 1834. Charred
MU

314 Subscribers lists to Dalton testimonial. 1833. 11 vols. 18 x 12 cms. to 16 x 10 cms.
MU

315 Draft of deed poll relating to statue of Dalton. 3 sheets. 40 x 25 cms.
Charred
MU

316 Certificate of admission as a burgess and guild brother. Edinburgh, 9th Sept. 1834. 1 sheet. 46 x 38 cms.
MU

317 Parchment certificate of honorary membership of the Royal Medical and Chirurgical Society. 16 Aug. 1836. 1 sheet. 48 x 37 cms.
MU

318 Solsom, Peter Harrogate well dispute papers, 1836–7
Endorsed 'Dr. Dalton coincides in opinion with Dr. Turner and the others who claim Mr. West's views untenable'.
MU

319 Daltonian Professorship and **Monument Committee** Minute book 26th Aug.–6th Nov. 1844. 20p.
MPL

320 Exors. of John Dalton Account of estate of John Dalton 1844–5. 18p. of which 3 blank. 20 x 16 cms.
MU

321 [Benson, William?] Pedigrees of Dr. John Dalton of Manchester and H. Dalton of Eaglesfield drawn up in order to ascertain whether both could be traced to the same common ancestor. 1 sheet. 25 x 40 cms.
MU

322 Benson, William Pedigree of John Dalton, D.C.L., LL.D., F.R.S., etc. of Manchester. 1847? 1 sheet. 54 x 65 cms.
Bears note 'copied from the pedigree in the possession of the Manchester Literary and Philosophical Society'
MU

323 Dalton relics formerly belonging to Dalton Hall. Typescripts 3 sheets. 33 x 20 cms.
Most of the items listed are now in the possession of either the Science Museum, the Museum of Science and Industry in Manchester, or Manchester University
LP

?● Letters to Dalton

See also the index of letters, pages 147–154

324 Bealey, John
Radcliffe, 15 Nov. 1838
MU

325 Bickerstaff, Robert
London, 31 Mar. 1818
Publisher of *A new system of chemical philosophy*
MU

326 Barker, E.H. 1788–1839, Classical scholar (D.N.B.)
London, 28 Feb. 1838; 20 Mar. 1838
Request for support of application for post of Registrar, University of London.
MU

327 Casaseca, J.L.
Madrid, 2 Jan. 1831
MU

328 Cuthbertson, John
London, 10 May 1814; 12 June 1814
MU

329 Davidson, James
Glasgow, 18 Dec. 1832
With copy letters from Barlow to Davidson and Davidson to Davy.
MU

329.3 Davy, Humphry
London, 1810
Managers of the Royal Institution hope to discharge their debt to Dalton. 'My estimation of phosphorated hydrogen does not accord with yours.'
Myers catalogue, 1977, no. 10. Lot 312

329.5 Davy, Humphry
London, 12 Feb. 1818
Asks if Dalton is interested in joining a polar expedition. Text in Thackray 492.5
RI

330 Dawson, Joseph
Royds Hall, 20 Jan. 1812

Reference ironstone specimen.
MU

330.3 Dickinson, Joseph
Maryport, 8 Feb. 1794
Colour blindness.
CPL

330.5 Dickinson, Joseph
Maryport, n.d.
Colour blindness.
CPL

331 Emmett, Anthony 1790–1872, Major-General, R.E. (D.N.B.)
Bermuda, 28 July 1838
MU

332 Giles, Francis 1787–1847, Civil engineer (D.N.B.)
London, 17 Feb. 1833; Gloucester, 23 Sept. 1836; London, 27 Nov. —.
MU

333 Howard, Luke 1772–1864, Quaker, meteorologist, chemist, apprenticed to Ollive Sims (D.N.B.)
—, 16 Jan. 1812; Pontefract, — 1831; Ackworth, 19 Sept. 1833; Tottenham, 1 June 1835; Manchester, 24 Aug. 1835; Ackworth, 24 Mar. 1837; 14 May 1842; 25 Jan. 1844
MU

334 Hunter, J.W.
London, 19 Mar. 1838
Request for support of application for post of Registrar, University of London.
MU

335 Johnstone, C.J.
London, 20 June, 1836
MU

336 See Suliot, T.E. 349.1

337 Lyall, Robert 1790–1831, Botanist and traveller (D.N.B.)
Moscow, 8 Oct. 1822
MU

338 Montgomery, J. President, Literary and Philosophical Society, Sheffield.
Sheffield, 5 Mar. 1824
MU

339 Otley, Jonathan
Keswick, report on Derwent Lake, 1830; 28 Nov. 1832; 11 Jan. 1838
Otley was the author of the first good guide to the Lake District and
Dalton's companion on many of his visits there.
MU

340 See Otley, Jonathan 355.5

341 Pease, Joseph 1799–1872, Railway projector and first Quaker M.P.
(D.N.B.)
— Oct. 1838
MU

342 Richardson, Charles 1775–1865, Lexicographer (D.N.B.)
Norwood, — Mar. 1838
MU

343 Sharpe, John, F.R.S.
London, 15 Nov. 1824; 7 Dec. 1826; —, 1826; 22 Sept. 1827; 9 Sept. 1828; 6
Mar. 1832
Vol. 2 of *A new system of chemical philosophy* is dedicated to Sharpe; the 1827
letter acknowledges this dedication.
MU

344 Sharpe, N. Mrs.
London, 30 Apr. 1844
MU

345 Sibson, Edmund
Warrington, 27 June 1828; 31 Oct. 1828; 12 Nov. 1828; 25 Feb. 1829; 23 Apr.
1829
MU

346 Sims, Ollive
Baltimore, 6 Mar. 1828
Reports delivery of a copy of *A new system of chemical philosophy* to the
American Philosophical Society where it is now in the Society's rare book
room.
MU

347 Sladen Charles
Manchester, 10 Apr. 1834
MU

348 Spencer, J.
Cockermouth, 6 Feb. 1838
MU

349 Spring-Rice, T. 1799–1866, 1st baron Monteagle. Secretary to the Treasury (D.N.B.)
London, 24 Nov. [1834?]
MU

349.1 Suliot, T.E.
Leeds, 3 Mar. 1837; 16 Mar. 1837
MU

350 Taylor, Charles
London, 17 Sept. 1838
MU

351 Watson, H.H. Instrument maker.
Bolton, 21 Feb. 1837; 9 Feb. 1842
Mentioned by Roscoe, *Proceedings*, 1876, **15**, 77–82.
MU

352 Watson, Thomas, Sir 1792–1882, Physician (D.N.B.)
London, 12 Nov. 1834
MU

353 Wettenhall, C.N.
Northwich, 1 Sept. 1818
MU

354 Whalley, Lawson
Lancaster, 11 Dec. 1833
MU

355 Yates, James 1789–1871, Unitarian and antiquary (D.N.B.)
Liverpool, 26 Oct. 1836
MU

✌ Letters about Dalton

See also the index of letters, pages 147–154

355.1 Daubeny, C.G.B., Oxford
to William Prout, 27 Oct. 1831 'You will find inserted in the appendix… some comments of Dalton's upon my exposition of his theory.' Text in Cardwell **478.2** p. 253. Bound in Prout's copy of Daubeny's *An introduction to the atomic theory* are Dalton's comments. (W.H. Brock)
RI

355.2 Forbes, J.D., Edinburgh
to Miss Forbes, Roehampton, 25 June 1831 'I spent several very happy days

at Manchester... The most extraordinary man I met was John Dalton whose name is almost better known in almost any country of Europe than his own & in any town of it than Manchester. He is usually styled by Continental writers the father of Modern Chemistry... Yet this man... is earning a penurious existence by teaching boys elementary mathematics...' Text in *Memoirs*, 1973–74, **116**, 10–1
UStA

355.3 Henry, W.C., Teignmouth
to John Davy, 14 Sept. [1853?]
Encloses letters from Humphry Davy to Dalton. Assessment of Dalton. Text in Thackray **492.5**
RI

355.4 Henry, William, Manchester
to C.G.B. Daubeny, Oxford, 30 Aug. 1831
Approves dedication of Daubeny's book on the Atomic Theory to Dalton 'in full confidence that he will be gratified by the compliment'. Discusses some uncertainties connected with the theory. Text in Patterson **487.5**
p. 227–30
Magdalen College, Cambridge. Reference: Daubeny Letter Book

355.5 Otley, Jonathan, Keswick
to Peter Clare, Manchester, 30 Aug. 1844
Appears to be a copy-letter acknowledging a copy of Dalton's will. 'Although the late Dr. Dalton always treated me as a companion, he would never permit me to go without some remuneration; I would not say for loss of time, as no time could ever be lost that was spent in his company; he was so affable and communicative. When on the last occasion I would have declined what he offered; he said *I must take it*; it might probably be the last and so it was, as far as regarded mountain excursions. Subsequent meetings were attended with a kind of melancholy pleasure.' Then follow two later notes by Otley:– 29 Nov. 1852. Sent the manuscript of memorandum of excursions with Dr. Dalton, 1812–1836, to Mr. Jas. Wooley, 69 Market St., Manchester. Recd. back 30 June 1853. 15 Aug. 1853. Sent copy of narrative to Dr. Henry, Haffield, near Ledbury. Henry **480** p. 147–60
KFM

355.6 Roberton, John, Manchester
correspondent unknown, 13 Sep. 1844
Description of Dalton by an eminent obstetrician and gynaecologist who was a member of the Lit and Phil and President of the Statistical Society. Text in Cardwell and Mottram, *Notes and Records Roy. Soc.* 1984, **39** (1), 29–40
Roberton Carver, Esq., Norwich: mf. UMIST

356 Jonathan and John Dalton, respectfully inform their friends, and the public in general, that they intend to continue the school lately taught by George Bewley... the school to be opened on the 28th of March, 1785. ... Kendal, printed by W. Pennington. Handbill, 14 x 16.5 cms.
Facsimile in Roscoe **488** and reproduced in Patterson **487.5**
SM: LP facsim.

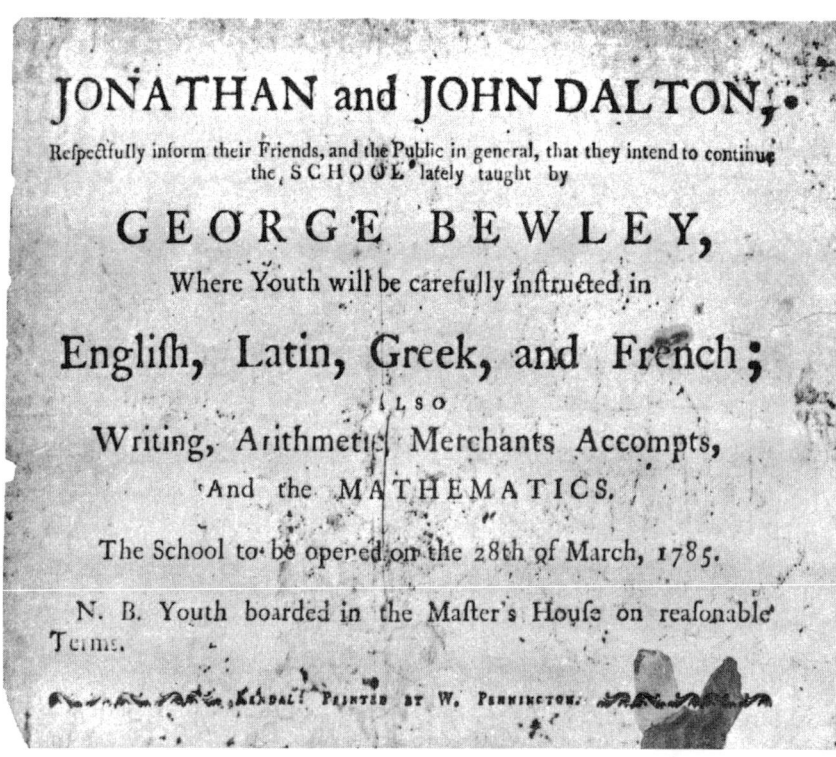

JONATHAN and JOHN DALTON,

Respectfully inform their Friends, and the Public in general, that they intend to continue the, SCHOOL lately taught by

GEORGE BEWLEY,

Where Youth will be carefully instructed in

English, Latin, Greek, and French;

ALSO

Writing, Arithmetic Merchants Accompts,

And the MATHEMATICS.

The School to be opened on the 28th of March, 1785.

N. B. Youth boarded in the Master's House on reasonable Terms.

KENDAL: PRINTED BY W. PENNINGTON.

A handbill for Jonathan and John Dalton's school, 1785 **356**

356.2 Kendal, July 5th, 1786. Jonathan and John Dalton take this method of returning their acknowledgements to their friends and the public for the encouragement they have received since their opening school and... are induced to hope for a continuation of their favours. ... Handbill
Reproduced in Roscoe **488** and Patterson **487.5**

356.4 Boarding School. Kendal, Jan. 16, 1787 Jonathan and John Dalton take this method of informing their friends and the public, that they have lately

taken and furnished a commodious house in Kendal,... with an intention of taking youth to board, and be instructed in all or any of the following branches of learning. ... *Cumberland Pacquet*, 24 January 1787. 10 x 7 cms.
CPL

356.5 [Kendal], Oct. 26, 1787. Twelve lectures on natural philosophy to be read at the school... by John Dalton. To begin on Tuesday evening, the 13th of November next... Handbill
Lectures: 1 and 2 Mechanics; 3, 4 and 5 Optics; 6 and 7 Pneumatics; 8, 9 and 10 Astronomy; 11 and 12 Use of Globes
A framed copy of this notice was on exhibition in the Society's House
The text is reproduced in Roscoe **488**

356.6 Kendal, Nov. 2, 1791 J. Dalton's philosophical lectures will begin on Monday evening the 14th of November at 6 o'clock, and continue on Mondays and Thursdays following. Kendal, printed by W. Pennington. Handbill, 24 x 18 cms.
Lectures: 1 and 2 Mechanics; 3 and 4 Optics; 5 and 6 Hydrostatics; 7 Fire; 8, 9 and 10 Astronomy; 11 and 12 Use of the Globes
Reproduced in Gee **378**

357 *Ladies' diary* 1789, 1794–5, 1797, 1818–19
Dalton's copies.
MU

358 *Gentleman's diary* 1793
Dalton's copy.
MU

359 Tracts from Dalton's library
VOL. 1. 1. The theory of the syphon. 1781 2. Stevenson, W. Remarks on the very inferior utility of classical learning. 1796. Inscribed 'The author to J. Dalton' 3. Barnes, T. Sermon preached at Rochdale. 1806 4. Davy, H. A lecture on the plan for improving the Royal Institution. Inscribed 'Mr. Dalton from the Author' 5. Hume, J. On the identity of silex and oxygen. 1808. Inscribed 'Mr. Dalton with respects from the author' 6. Dalton, R. Syllabus of a course of lectures on natural philosophy. Inscribed 'Syllabus of R. Dalton's Lects. on Natl. Philosophy Manch. 1806 (John Dalton's copy)'
VOL. 2. 1. Simmons, W. A detection of the fallacy of Dr. Hull's defence of the cesarean operation 2. Simons, W. Reflections on the cesarean operation. Inscribed 'Mr. Dalton with Mr. Simmons's respectful compts.' 3. An address to the proprietors and managers of coal mines on destroying the fire-damp. 1806 4. A reply to Dr. Trotter's second pamphlet respecting fire-damp. 1806 5. Dewar, H. Letter to Thomas Trotter
MPL

J. DALTON'S

PHILOSOPHICAL LECTURES

WILL BEGIN

On Monday Evening the 14th of November, at 6 o'Clock, and continue on Mondays and Thurfdays following.

Admittance Sixpence each Lecture, or five Shillings the whole.

⟶⟶⟶⟶⟶✦✦❯❮✦❮⟵⟵⟵⟵⟵

LECTURES I. & II. MECHANICS.

On Matter, and its Properties. Attraction and Repulfion in general Experiments upon electric, chymical, and particularly magnetic Attraction and Repulfion. On Gravity. The Laws of Motion, Mechanic Powers. Vibration of Pendnlums.

LECTURES III. & IV. OPTICS.

Nature and Properties of Light. Simple Vifion. Doctrine of Colours. Of reflected Vifion, Mirrours, &c. Of refracted Vifion, Lenfes, &c. Burning Glaffes. Defcription of the Eye, and manner of Vifion. Of Optic Glaffes. The Rainbow explained.

LECTURES V. & VI. HYDROSTATICS & PNEUMATICS.

Of Fluids in general. Properties of elaftic Fluids. Specific Gravity of Bodies. Of the Atmofphere. Defcription of the Air-Pump A great variety of Fxperiments on the Air-Pump, proving the fpring, weight, and other Properties of the Air. Account of Difcoveries upon fictitious Air, and common Air injured by Refpiration, Putrefaction, Combuftion, &c.

LECTURE VII. ON FIRE.

The Thermometer: Difcoveries relative to Heat confequent thereto. Of the fource of Animal Heat, and the nature of Combuftion.

LECTURES VIII. IX. & X. ASTRONOMY.

Of the Solar Syftem, and Syftem of the Univerfe. Various aftronomical Phenomena as the Phafes of the Moon, Eclipfes, Occultations, Tranfits, &c. explained. Of Tides.

LECTURES XI. & XII. USE OF THE GLOBES.

Defcription of the Globes, and a variety of Problems performed, and Phenomena illuftrated by them.

⟿⟿⟿⟿ KENDAL: PRINTED BY W. PENNINGTON. ⟿⟿⟿⟿

A handbill for Dalton's philosophical lectures, Kendal, 1791 **356.6**

359.1 Mathematical Academy, Manchester
John Dalton, Secretary to the Literary and Philosophical Society, and late
Mathematical Tutor at the New College, Manchester, respectfully informs
his Friends and the Public, that he intends shortly to open a
MATHEMATICAL and PHILOSOPHICAL ACADEMY in Manchester...
August 30, 1800
Advertisement in the *Manchester Mercury*, 2 September 1800
Illustration in Thackray **492.5**
MPL

360 Printed syllabuses of lectures, 1804–35
Constitution of mixed gases. Royal Institution, 23 Jan. 1804. 1 sheet 16 x 11
cms.
Programme covering two weeks of lectures by Dalton, Allen and Davy.
SM
Natural philosophy. Manchester , 1805. 8p. 22 x 14 cms.
MU
Heat and chemical elements. Edinburgh, 1807. 1 sheet 22 x 14 cms.
MU
Experimental philosophy. Manchester, 1808. 24p. 22 x 14 cms.
MPL; MU
Experimental philosophy and chemistry. Manchester, 1811. 8p. 22 x 14 cms.
MU
Mechanics. Manchester, 1818. 1 sheet 28 x 22 cms.
MU
Pharmaceutical chemistry. Manchester, 1824. 1 sheet 22 x 13 cms.
MU
Meteorology. Manchester, 1825. 1 sheet 23 x 19 cms.
MU
Meteorology. Birmingham, 1825. 1 sheet 23 x 19 cms.
MU
Chemistry. Manchester, 1827, 1828, 1829. 3 sheets 25 x 20 cms.
MU
Meteorology. Manchester, 1835. 1 sheet 26 x 20 cms.
MU

360.2 Banks, John Experiments on the velocity of air issuing out of a
vessel in different circumstances; with a description of an instrument to
measure the force of a blast in bellows, etc. Communicated by Mr. Dalton,
30 May 1800 *Memoirs*, 1802, **5**, 398–406
For the influence of Banks on Dalton, see Thackray **492.5** p. 44–6

360.5 Manchester Literary and Philosophical Society A list of members
of the Society with the date of their election, March 1827. Broadside 32 x 20
cms.

Dalton is shown as President, an office he held from 1817 to 1844.
Subsequent rules limited the period of Presidency
LP

361 The Dalton Testimonial. Notice and list of subscribers, [1833?] 1 sheet
of 4p.
MU

362 Rice, T.S. Speech of the Right Honourable Thomas Spring Rice... on
the pension list in the House of Commons on Tuesday, 18th of February,
1834
Extracted from the *Mirror of Parliament* – part 259. London, 1834. 17p. 21 x 13
cms.
Inscribed 'Mr. Spring Rice, speech in the Ho. of Coms. on the Pension List
London, March 3 1834. Dr. Bardsley, Ardwick, Manchester. Geo. W. Wood'.
Charred
MU

362.3 Dr. Dalton's Proposed Atomic Symbols Proceedings of the fifth
meeting of the British Association for the Advancement of Science, Dublin,
1835, – Appendix. Lithograph.
The table differs considerably from the one distributed at his Manchester
lecture in October, 1835. *See* Thackray **492.5** p. 117–9 where it is reproduced.
BL

362.4 Atomic Symbols by John Dalton, DCL, FRS, &c, &c,
explanatory of a lecture given by him to the members of the Manchester
Mechanics' Institution, 19th October 1835.
Lithographed for the Directors by T. Physick, King St. 27.5 x 21 cms.
Conjugate with a blank leaf of the same size.
Following the invitation of the Directors of the Manchester Mechanics'
Institution, Dalton gave a course of five lectures on meteorology beginning
in March 1835. 'Later in the year, Dalton gave a lecture at the Institution on
the Atomic Theory. To the audience was distributed a lithographed sheet of
atomic symbols. ...the lecture-room was crowded in every part and the
greatest anxiety was manifested by the audience not to lose a single word
which fell from the lips of the speaker.' – Gee **378**. It was his last public
lecture.
The table is included in the biographies by Henry (1854), Lonsdale (1874)
and Roscoe (1895); each of the three reprints is separately redrawn but none
includes the information about the lithographer. The Roscoe version
appears to have been widely distributed as a separate *See also* **75** and **378**
Roy G. Neville Historical Chemical Library, Redwood City, Calif.

363 Alley, P.B. Extracts from the *Manchester Guardian* and other documents
relative to the death and funeral of the late John Dalton. 1844

ATOMIC SYMBOLS

BY

John Dalton, D.C.L., F.R.S. &c. &c.

explanatory of a

LECTURE

given by him to the MEMBERS of the

Manchester Mechanics' Institution,

October 19th 1835.

ELEMENTS.

Hydrogen	Oxygen	Azote	Chlorine
Carbon	Phosphorus	Sulphur	Lead
Zinc	Iron	Tin	Copper

OXIDES.

SULPHURETS.

COMPOUNDS.

Binary.

Water
Nitrous gas
Carbonic oxide
Sulphuretted hydrogen
Phosphuretted hydrogen
Olefiant gas
Cyanogen

Ternary.

Deutoxide of hydrogen
Sulphurous acid
Acetic acid
Nitrous oxide
Carbonic acid
Phosphoric acid
Nitrous vapour
Carburetted hydrogen
Prussic acid
Bicarburetted hydrogen
Tan

Quaternary.

Sulphuric acid
Binolefiant gas
Pyroxylic spirit

Quinquenary.

Ammonia
Nitrous acid
Prussic acid

Sexenary.

Alcohol
Pyroacetic spirit

Septenary.

Nitric acid

Decenary.

Ether

An illustration of atomic symbols, distributed at a lecture in 1835
(from a lithograph of 1854) **362.4**

Scrapbook containing the following:
Newspaper clippings, July, 1844–February, 1853
Funeral notices and placards issued by the Committee of Management
and by the Manchester Corporation
Programme of the funeral procession
Notice of arrangements for the funeral for members of the Literary and

PUBLIC NOTICE.

By virtue of the power given to me in that behalf by an Act passed in the present Session, intituled "An Act for the Good Government and Police Regulation of the Borough of Manchester," I do hereby order and declare that the following Streets shall be freed from Obstruction between the hours of 10 and 1 o'clock, on MONDAY, the 12th day of AUGUST inst., so as to :,How the

PUBLIC PROCESSION,

On occasion of the Funeral of the late Dr.

DALTON,

To proceed uninterruptedly from the Town Hall, through the said Streets, viz:—Cross Street, St. Ann's Street, St. Ann's Square, Exchange Street, Market Street, Piccadilly, London Road, Ardwick Green, and Hyde Road, to the Ardwick Cemetery.

And I further order and direct that during the continuance of the Procession, all Officers of the Police shall preserve order, and prevent and remove all obstructions in the neighbourhood of the Town Hall and in the line of Procession.

I take this opportunity of intimating to my fellow townsmen, that by closing their Warehouses and Shops from 11 until 1 o'clock on Monday, they will best show their respect for the memory of the late Dr. DALTON.

ALEX. KAY,

Dated at the Town Hall,
This 10th day of August, 1844.   MAYOR.

Notice of Dalton's funeral **363**

Philosophical Society

Notice of arrangements for the funeral for the 'steam engine and machine makers, millwrights and others of the trade'

Details of inscription on coffin

Notice convening public meeting about Dalton Monument, 3rd October, 1844

Report of the proceedings of a public meeting of the inhabitants of Manchester, held to determine on a suitable memorial of respect for the memory of the late John Dalton. October 4th, 1844. 12p.

Notice of Committee meeting on 14th October, 1844, with resolution on Daltonian Professorship dated 4th October, 1844

Daltonian Professorship and Monument Committee, proposed scheme dated 14th October, 1844

Scheme for the management of the Daltonian Professorship as adopted by the Provisional Committee, 30th October, 1844

Daltonian Professorship Committee. Draft of proposed address to the public, 30th November, 1844

MPL

364 Hypatia

The lament on the death of the late John Dalton, D.C.L., F.R.S., etc. 1 sheet 4p. 18 x 11 cms.

Inscribed 'Mrs. Robert Abbatt with the respects of Hypatia'

MU

365 Hypatia

On the death of the late John Dalton, D.C.L., F.R.S., etc. 1 sheet 4p. 18 x 11 cms.

Inscribed 'Mrs. Robert Abbatt with the respects of Hypatia'

MU

366 Pall bearers

Inscription on Dalton's coffin. 1 card 4 x 11 cms.

MU

366.1 Friends Meeting House, Manchester

Burial certificate signed by Peter Clare 1844. 1 sheet 17 x 19 cms.

Entry in the Burial Notes of Manchester Meeting (p. 104). It replaces the usual printed form of instruction to the grave-digger with the necessary details in writing i.e. name, etc. Hence it is quite clear that although (for reasons outside the control of the Manchester meeting) Dalton was not buried in the Friends' burial ground, yet his burial, i.e. that of a member of the Meeting, was properly recorded. It is not in the minutes, however. – J.T. Marsh

Manchester Meeting: Friends Meeting House, Manchester.

367 The late Dr. Dalton's effects: catalogue of elegant household furniture... books, &c., which will be sold by auction on Thursday and Friday, Oct. 10 and 11, 1844. Salford, printed by William Jackson [1844] 16p. 20 x 12 cms.
MPL

368 Manchester Mechanics' Institution
Syllabus of two lectures on some of the researches and discoveries of Dr. Dalton... by Mr. Henry Day, jun. 1844. 6p. 21 x 13 cms.
MPL

368.5 Manchester, Joseph Letter to John Harland. Manchester, 17 March 1853. 1p.
With this you will receive an old thermometer which formerly belonged to the late Dr Dalton and is the one he had placed outside his window in Faulkner St for many years; it was purchased at the sale of the late Peter Clare's effects
LP

369 Woodcroft, Bennet Dalton's meteorologist. This table arranged by Bennet Woodcroft exhibits at one view the mean monthly state of the rain gauge, the barometer and the thermometer at Manchester from data kept for upwards of twenty years and tables formed by the late Dr. J. Dalton. 1854. 1 sheet 34 x 45 cms.
BL, MPL

370 Wilkinson, T.T. An account of the early mathematical and philosophical writings of the late Dr. Dalton *Memoirs*, 1855, **17**, 1–30
Describes Dalton's contributions to the *Ladies' Diary* and the *Gentleman's Diary*, 1783–95. See **22** and **23**

371 Baxendell, Joseph On a thermometer constructed by Dr. Dalton *Proceedings*, 1865, **4**, 133

372 Roscoe, Sir H.E. Notes on a collection of apparatus employed by Dr. Dalton in his researches which is about to be exhibited (by the Council of the Literary and Philosophical Society of Manchester) at the loan exhibition of scientific apparatus at South Kensington *Proceedings*, 1876, **15**, 77–82

373 Dancer, J.B. Note on the daguerreotype portrait taken of the late Dr. Dalton *Proceedings*, 1877–8, **17**, 43–4

374 Owens College, Manchester
On the occasion of the celebration of the centenary of Dalton's atomic theory, 1803: physiological exhibits by Professor Wm. Stirling, M.D., D.Sc. 1 sheet of 4 pages 22 x 14 cms.

Designed to show some interesting facts in connection with the physiology of vision
LP, MPL

375 Owens College, Manchester
Conversazione on the occasion of the celebration of the centenary of Dalton's atomic theory, 20th May, 1903. Programme. 1 sheet of 4p. 25 x 19 cms.
LP

376 Jones, Francis The collection of apparatus used by Dalton, now in the possession of the... Society *Memoirs*, 1904, **48** (22), 5p.
Includes 9 plates of photographs of Dalton's apparatus to which the text refers. Some of this apparatus is in the Museum of Science and Industry in Manchester

377 Cohen, Ernst Een physisch-chemische Caricatuur *Chem. Weekbl.*, 1905, **2**, 97–111 *Mitt. Gesch. Med. Naturw.*, 1905, **4** (15) 253–70
Includes reproduction of caricature of Dalton thanking Gerrit Moll for defending English science. Artist unknown but possibly Thackeray. See **461**

378 Gee, W.W.H., Coward, H.F. and **Harden, A.** John Dalton's lectures and lecture illustrations *Memoirs*, 1914–15, **59**, (12) 66p
1. Gee, W.W.H. John Dalton's lectures. Details of lectures, 1787–1835, with illustrations of syllabus, 1791, and Dalton's ledger, 1791. 2. Gee, W.W.H. John Dalton's natural philosophy diagrams. 3. Coward, H.F., and Harden, A. The lecture sheets illustrating the atomic theory. 19 sheets are illustrated.

379 Barnes, C.L. The Society's house *Proceedings*, 1915–16, **60**, 2–8
Written for the British Association meeting, 1915. Lists the Dalton items in the possession of the Society at that time

380 Adamson, R.S., and **Crabtree, A., McK.** The herbarium of John Dalton *Memoirs*, 1918–19, **63** (1), 46p.

381 Hickson, S.J. Some early autographs of John Dalton *Proceedings*, 1921–2, **66**, 4–5
Describes Dalton's ms. additions to the copy of *The Schoolmaster's Assistant* presented to the Society by Professor Hickson. See **141**

382 Some Dalton letters *Manchester City News,* 18 November, 1922. Letters to Jonathan Dalton, 1792, 1799, 1805 and 1817. *Proceedings*, 1922–3, **67**, 9–10
See also **167.22–167.28**

383 Gee, W.W.H. Four letters of John Dalton *Proceedings*, 1922–3, **67**, 9–10
Describes four letters to Jonathan Dalton dated 1792, 1799, 1805 and 1817, presented to the Society by Colonel Allen. See also **167.22–167.28**

384 Gee, W.W.H. John Dalton's spectacles. John Dalton's pupil's note-book *Proceedings*, 1925–6, **70**, 6–8
Describes John Dalton's spectacles and the note-book of one of his pupils, presented to the Society by the Misses Taylor

384.5 Marsden, Arthur 'Visited 36, George St., Manchester, in July 1929. Tried on Dalton's hat. Read his diary and handled the collection… Stayed at

ELEMENTS

		w.t			w.t
⊙	Hydrogen.	1	✛	Strontian	46
◐	Azote	5	✸	Barytes	68
●	Carbon	54	Ⓘ	Iron	50
○	Oxygen	7	Ⓩ	Zinc	56
⊗	Phosphorus	9	Ⓒ	Copper	56
⊕	Sulphur	13	Ⓛ	Lead	90
◉	Magnesia	20	Ⓢ	Silver	190
⊖	Lime	24	ⓓ	Gold	190
◫	Soda	28	Ⓟ	Platina	190
⦿	Potash	42	✺	Mercury	167

A lecture sheet used by Dalton **378**

Dalton Hall where they had in the library Dalton's umbrella, cradle and his eye. The latter shamefully kept, as all the spirit had evaporated. Visited Schunck Library (Vic. Univ. of Manc.) and saw Dalton's chair and Davy's balance.'
Inscription in LP copy of Meldrum **483**

385 Garnett, Henry Photographs of John Dalton. Letter. *Nature*, 1931, **127** (3197), 201.
Letter asks for whereabouts of 3 daguerreotypes taken of Dalton. *See* **455**

385.2 Manchester Literary and Philosophical Society *The Thomas Thorp room* February 1936, 2p., 29cm.
The new room was added to the first floor of the Society's old House in 1936 as a memorial to Thomas Thorp, a noted inventor. It met the need to display the large collection of scientific apparatus and other relics accumulated since 1781. 'Of Dalton relics the Society possesses what might be called a complete collection, including the wall diagrams drawn by his own hand to elucidate the atomic theory, a quantity of very primitive-looking glass apparatus, thermometers, wooden blocks, balance and weights, tin bottles with glass tubes luted into their tops (india-rubber stoppers being then unknown), a small single action air-pump, and a ponderous one worked by a revolving handle...'
LP

386 Manchester Literary and Philosophical Society List of articles salvaged from 36, George Street, Manchester, after the destruction of the building on 24th December, 1940 *Proceedings*, 1939–41, **84**, 34–6

387 Manchester Literary and Philosophical Society List of some of the medals awarded to John Dalton and others which are the property of the Society. *Proceedings*, 1939–41, **84**, 31–3. See also **1034–1037**

388 Watson, E.C. Caricature of Dalton thanking Gerrit Moll for defending English science *Amer. J. Phys.*, 1946, **14** (1), 54
Includes reproduction of caricature and is based on Brockbank's and Cohen's articles. The article coincided with the announcement of Dr. Cohen's death at Auschwitz. See **461**

389 Duveen, D.I. and Klickstein, H.S. John Dalton's autobiography *J. Chem. Ed.*, 1955, **32**, 333–4
A brief note, dated 27th October, 1832, by Dalton on his life to Richard Roberts, included in the latter's 'Book of Autographs' (not published, ms. copy in possession of D.I. Duveen). Almost identical with one inscribed to Miss Johns, written 19th February, 1833 and included in Henry's and Roscoe's biographies

390 Neville, R.G. Unrecorded Daltoniana: two letters to John Bostock and a prospectus to the *New System*, 1808 *Ambix*, 1960, **8** (1) 42–5
Includes illustrations of the prospectus which with two letters to Dr. John Bostock, Liverpool, (dated 11th and 21st June, 1808) is in author's possession

391 Brockbank, William Dalton pedigree correspondence. 1961.
Typescript
MPL

392 Ardern, L.L. A unique John Dalton photograph? *Memoirs*, 1961–2, **104**, 67–9
Includes enlargement from the 1842 daguerreotype. See **455**

392.1 Manchester Education Committee The John Dalton College of Technology: programme of the official opening on 19 December 1964 by the Prime Minister, the Right Honourable Harold Wilson, OBE, MP [1964], 8p.
The College has now become part of the Manchester Metropolitan University.
LP

392.15 Dalton (Eaglesfield) MSS. Calendar of mss. deposited in the Cumberland Record Office by Mr John Dalton, 11 May 1965. 4p. typescript
Lists 40 groups of material dating from 1592
LP

392.2 Manchester Museum [Manchester University] *The Dalton exhibition*, 1966. Manchester, Manchester Museum, 1966 Exhibition to mark the bicentenary of the birth of John Dalton, scientist, held in Manchester Museum, 19th September to 28th October, 1966. Introduction by Horace Hayhurst. 8p.
Lists the individual exhibits and indicates ownership of each
LP

392.3 Japan Academy Address of homage to the memory of John Dalton, on the occasion of the 200th anniversary of his birth, from the Japan Academy. 17p. ms. in Japanese script, 6p. English translation in typescript
Signed by Dr Yuji Shibita, President, the Japanese Academy, 19 September, 1966
LP

392.4 Manchester University John Dalton bicentenary celebrations: reception and degree ceremony, 22 September 1966. Presentation by Prof E.G. Rupp of Prof Herbert Butterfield, Prof Michael Polanyi and Prof Harold Clayton Urey. Typescript, 5p.
LP

392.5 Manchester Literary and Philosophical Society Scrapbook relating to the Dalton bicentenary celebrations, 1966
Includes programmes, invitation cards, newspaper clippings, photographs, menus, guest lists and other miscellaneous items
LP

392.55 Manchester Museum of Science and Industry Manchester Literary and Philosophical Society, John Dalton Collection. [1981] 2p. typescript
Details apparatus and other items in the Museum
LP

392.6 Manchester Literary and Philosophical Society Correspondence and other documents relating to the valuation and sale of the Dalton mss. to the John Rylands University Library of Manchester. 1979
LP

392.65 *Dalton Tradition News* Dalton's papers saved. Manchester, John Rylands University Library, 1991 broadsheet, 2p., 42cms. Produced in conjunction with the exhibition, 'The Dalton tradition', held in the Library, 16 April–26 June 1991. Describes the work of the Library's conservation staff in preserving the Dalton papers. See also Leitch **481.7**

392.7 *The John Dalton Conference*: an international conference celebrating 200 years of colour vision research, 9–13 September 1994, Manchester. Organised by the Visual Sciences Laboratory, UMIST
1. Programme and information for delegates. 10p.
2. The John Dalton Lecture, The Library Theatre, Manchester, 11 September 1994.
   Programme–
   20 February 1794 Dear Cousin…
   31 October 1794 Inaugural lecture to the Society (John Dalton played by Malcolm Raeburn)
   200 years later
   Dr David Hunt. The genetic analysis of John Dalton's eyes
   Dr John Mollon. The link between the genotype and phenotype of Daltonism
LP See also **993**

❧ Minor references in *Memoirs and Proceedings*

---

393 Wooley, James Memoir of Dalton from the journal of the late Jonathan Otley [reference] *Proceedings*, 1860, **1**, 50

Both Angus Smith and Henry used this memoir in the preparation of their respective biographies

394 Preservation of John Dalton's apparatus *Proceedings*, 1860, **1**, 53–4

395 Gift – Dr. Dalton's manuscript correspondence – Dr. W.C. Henry *Proceedings*, 1864, **3**, 57

396 Return of Dalton's apparatus and instruments from the loan exhibition South Kensington and future preservation *Proceedings*, 1877, **16**, 228

397 Gift – Photographic portrait of Dr. Dalton enlarged from a daguerreotype. Photograph of the first prospectus of Dalton's school – Mr. G.S. Wooley *Proceedings*, 1877–8, **17**, 27

398 Faraday, F.J. List of interesting gifts to the Society including Dalton's apparatus and mss. by W.C. Henry, 1863, and a bust of Dalton by Mrs. Samuel Fletcher in 1864 *Proceedings*, 1892, **35**, 2

399 Gift – two receipts for private lessons, 24 June and 29 September, 1831 – Mr. George Bowman *Proceedings*, 1896, **40**, 99 and 108

400 Gift – Cabinet in which to preserve the Dalton manuscripts – Sir Henry Roscoe *Proceedings*, 1896, **40**, 108

401 Exhibit of oil painting of Dalton [no further details] *Proceedings*, 1896, **40**, 58

402 Restoration of Dalton's tomb. Gift – Dalton's diploma of honorary membership of the Edinburgh Medical Society *Proceedings*, 1899, **44**, 2

403 Dalton's tomb. Illustration of the tomb-frontispiece. *Proceedings*, 1899, **44**, 29 and 50

404 Gift – Dalton's 'Grammar' – Mr. W.E. Hoyle *Proceedings*, 1901, **45**, 21

405 Gift – Mural tablet in white marble, placed in the Secretaries' room, bearing the following inscription: 'This room was the laboratory of John Dalton: here his great discoveries were made, and here he conceived and worked out his atomic theory' – Dr. Edward Schunck *Proceedings*, 1901–2, **46**, 51

406 Letter from D. Mendeléeff referring to Dalton *Proceedings*, 1903–4, **48**, 28

407 Gifts: – Silhouette portrait – Mr. A. McDougall; engraved portrait by J. Stephenson – Mr. Charles Leigh: marble columnar pedestal for the bust of Dalton – Dr. Henry Wilde *Proceedings*, 1903–4, **48**, 27

408 Unveiling of bronze bust of Dalton, the work of Miss Levick, and based on the bust executed by Chantrey in 1833 and a statuette presented to the Society by Dr. Bealey. Presented by Sir Henry Roscoe as a memento of his years of association with the Society. Illustration of bust – frontispiece *Proceedings*, 1903–4, **48**, 3. See also **469**

409 Gift – Volume of scientific memoirs which had been in the possession of Dalton, Hodgkinson and Fairbairn – Mr. F. Nicholson *Proceedings*, 1905–6, **50**, 21

410 Gift – Framed enlarged photograph of the statues of Dr. Dalton and Dr. Joule in Manchester Town Hall – Sir W.H. Bailey *Proceedings*, 1906–7, **51**, 7

411 Gifts – Engraved portrait and a lock of Dalton's hair – Dr. F.W. Jordan *Proceedings*, 1907–8, **52**, 2

412 Reply to request for loan of Dalton's apparatus from the Committee of the Franco-British Exhibition *Proceedings*, 1907–8, **52**, 35

413 Exhibition of drawing of John Dalton signed by J. Stephenson and thought to be the original from which the celebrated engraving was made *Proceedings*, 1908–9, **53**, 23

414 Gift – Letter to Elihu Robinson, dated '2nd mo. 20th, 1794' – Mr. Francis Nicholson *Proceedings*, 1911–12 **56**, 38

415 Gift – Microscope which formerly belonged to Dalton – Dr. A.E. Barclay and Mr. John Noton *Proceedings*, 1916–17, **61**, 26

416 Gift – Original painting by Ford Madox Brown for the fresco in Manchester Town Hall of Dalton collecting marsh gas – Mr. Henry Boddington *Proceedings*, 1917–18, **62**, 2

417 Gift – Large medallion of Dalton – Mr. P.E.B. Jourdain *Proceedings*, 1918–19, **63**, 2

418 Gifts – Marriage certificate witnessed by Joseph Dalton; another witnessed by John Dalton; a copy of *The Schoolmaster's Assistant* which contains Dalton's signature inscribed when he was nine years old – Mrs. S.J. Hickson and Mrs. L.H. Fletcher *Proceedings*, 1921–2, **66**, 33

419 Gifts – Silver medal of Dalton, 1842; copies of the *Illustrated London News*, 1844 *Proceedings*, 1924–5, **69**, 31

420 Gift – John Dalton's pupil's notebook – the Misses Taylor *Proceedings*, 1925–6, **70**, 7

421 Joule, J.P. Autobiographical note *Memoirs*, 1930–1, **75**, 109–15

422 Gift – Air pump believed used by Dalton – Mr. A. McCandlish *Proceedings*, 1932–3, **77**, 20

423 Gifts – Gold watch given to John Dalton by Peter Clare – Miss A. Wood; piece of glass apparatus – Major Peter Groves; meteorological register *Proceedings*, 1934–5, **79**, 4–5

424 Gifts – Miniature – Mr. E. Goodhew; daguerreotype and letter – Miss Law; portrait engraving – Mr. H. Hampson; small portrait engraving – Miss Rooke; photograph of grave – Dr. Ashworth; thermometer – Mrs. K. Smithells; photograph of a certificate – Mr. J.H. Walmsley; 'Meteorology' – Miss K. Dornan *Proceedings*, 1943–5, **86**, 14

425 John Dalton's grave renovated *Proceedings*, 1945–6, **87**, 8

426 Replica of the portrait of Dalton, hung in Central Library *Proceedings*, 1946–7, **88**, 4

427 Gift – Engraving of Dalton – Mr. Brayshaw *Proceedings*, 1947–8, **89**, 6

428 Council has agreed for certain relics to be sent to the South Kensington Museum on loan. Gifts – Four early volumes of John Dalton – Royal Agricultural Society; portrait and four engravings – Miss D. Wood; 'Phosphates and Arseniates' – Miss Chadwick *Proceedings*, 1948–9, **90**, 12

429 Gift – Print of John Dalton – General Sir H. Jackson *Proceedings*, 1951–2, **93**, 7

430 Gifts – Letters and mementoes – Mr. Taylor; tables – Prof. Travers *Proceedings*, 1956–7, **98**, 3

431 Gift – Bust of Dalton – Dr. Petch *Proceedings*, 1961–2, **104**, 5

432 Gift – ms. of a paper signed by Dalton, dated September 14th, 1801, with the title 'New theory of the constitution of mixed aeriform fluids and particularly of the atmosphere' – Mr. H.C. Wilson *Proceedings*, 1962–3, **105**, 5

432.1 Gift – Certificate of membership of the Society of the Revd. J.R. Beard, dated 1827 and bearing the signature of John Dalton – Mr M.H. Winder
Thanks to Dr A.P. Hatton for his assistance in having arranged the renovation and mounting of a model steam engine made by Clegg, one of John Dalton's pupils *Proceedings*, 1963–4, **106**, 4–5

432.2 The Dalton mss. which survived the destruction of the Society's library in 1940, have been sorted and catalogued during the year. The publications sub-committee were concerned with the arrangements for the publication of a bibliography of John Dalton to be issued before the celebration of the Dalton Bicentenary in September 1966
Gift – Seven microfilm spools of books and periodical articles relating to Dalton – USSR Academy of Sciences *Proceedings* 1964–5, **107**, 4–5

432.3 Microfilms have been made of the surviving items of the Society's collection of Dalton manuscripts; these are in such a charred and brittle condition that their future preservation is problematical
During the session the Dalton bibliography was published by the

Manchester University Press in good time for the bicentenary of his birth in September 1966; the Society is shown as the owner of the copyright
Film sequences were taken in the Society's House on November 22nd 1965 for inclusion in a film on the life of Dalton and his work, being prepared by ICI Ltd. **472.2** *Proceedings*, 1965–6, **108**, 3–5

432.4 Report on the Dalton bicentenary celebrations
The Dalton manuscripts which were deposited on loan in the Science Museum, South Kensington, in 1948 have now been returned. *Proceedings*, 1966–7, **109**, 3–4 and 6–7

432.5 A suggestion was put forward by the Council that a plaque should be inserted in the wall of the new Bank of England building recording the fact that this would extend over the site of John Dalton's house in Faulkner Street. This suggestion was accepted by the Bank and by the Manchester Corporation
Gift – Four photographic transparencies of John Dalton's birthplace at Eaglesfield – Mr J.T. Marsh *Proceedings* 1967–8, **110**, 6–7

432.6 Gift – Two photographic plates of John Dalton – The Manchester Geographical Society *Proceedings*, 1970–1, **113**, 9

432.7 At a General Meeting of the Society on October 30th, 1972, Dr Arnold W. Thackray who has recently published a biography of John Dalton, was elected a Corresponding Member. In acknowledging his Certificate of Election, Dr Thackray pointed out that a connection with Philadelphia and the University of Pennsylvania is not without its historical precedent for the Society. Benjamin Rush, Joseph Priestley and James Mease MD were all members of the Society.
Gift – *John Dalton: a critical assessment of his life and work* – the author, Dr A.W. Thackray *Proceedings*, 1972–3, **115**, 3 and 8

432.8 On 11 July 1979 the Society was honoured by a visit by the Duke of Edinburgh to unveil a plaque to John Dalton. His Highness also showed considerable interest in the exhibition of Daltonalia which had been mounted in the library.
The surviving Dalton manuscripts were sold to the Manchester University Library
The Curator has supervised the transfer of the apparatus and exhibits to the North West Museum of Science and Industry, to be held there on loan *Proceedings*, 1977–80, **120**, 5

432.9 One happy discovery in the process of demolition of the Society's House was the uncovering of a bronze bust of John Dalton, presented by Sir Henry Roscoe. The developers, French Kier Holdings Ltd., restored the bust, finally presenting it to the Society *Proceedings*, 1981–2, **122**, 1

‰ Portraits, sculptures, film and television, illustrations

Portraits

There are three main sources for the portraits of Dalton: the Allen painting, 1814; the 1834 drawings made when Dalton visited London in that year; the Nicklin daguerreotypes, 1842

433 Allen, Joseph Oil on canvas. 1814.
Presented to the Society in 1816 and destroyed in 1940. See also **310**

434 Allen, Joseph Oil on canvas. 1946. 135 x 100 cms.
Replica by Maud Hall Neale. This portrait together with that of Joule is hung in the Society's office
LP

435 Allen, Joseph Oil on canvas (quarter size). Date unknown. 70 x 58 cms.
It is not known if this portrait is by Allen or merely a copy. There are slight differences from the larger portrait. The right hand is thrust into the waistcoat and the colour of the coat is brown and not dark grey. Presented by Thomas Agnew to Salford City Art Gallery, 1868
UMIST

436 Allen, Joseph Engraving. Artist: Joseph Allen. Engraver: Worthington. Publisher: Agnew and Zanetti. June, 1823. 36 x 28 cms. (See frontispiece)
Some copies are dated December, 1834

437 Allen, Joseph Engraving. Artist: Joseph Allen. Engraver: Cooke. Publisher: W. Mackenzie. n.d. 15 x 13 cms.
MPL Reference: 6559: M618/5

438 Allen, Joseph Engraving. Artist: Joseph Allen. Engraver: W. Morton. 1856. 19 x 13 cms.
Frontispiece in Smith's 'Memoir'. Under the engraving is written 'The weather is awkward; but the glass is rising. My best respects to all. Yours truly, John Dalton'.

439 Baker, I.H. Engraving of head. 1854. 20 x 11 cms.
From the statue by Chantrey. Frontispiece in Henry's *Memoirs* **480**

439.2 Bradley, John Oil on canvas. Date unknown but probably not from life. 99 x 58 cms. oval
Reproduced in Cardwell **478.2**
MU Dalton-Ellis Hall

439.5 Bradley, William Crayon drawing. Dalton in old age. 32 x 25 cms.
Lent by Mr W.M. Lord of Todmorden to the Dalton Exhibition,
Manchester, 1966

440 Brockedon, William Pencil and wash. 1834
Dalton stated in a letter to the artist 'It is the best that I have seen'.
Brockedon in later years devoted himself to science rather than art and took
out a number of patents. He was founder of the Graphic Society
NPG Reference: 2515/66 See also **162** and **925**

441 Brown, F. Madox Dalton collecting marsh fire gas. Oil painting. 1886
Study in oils, presented to the Society in 1917 by Mr. Henry Boddington and
destroyed by enemy action, 1940

442 Brown, F. Madox Dalton collecting marsh fire gas. Mural (toiles
marouflées). 1887. 300 x 150 cms.
'The head of Dalton was painted from a little portrait bust in the possession
of Mr. Joule of Sale'
Manchester Town Hall

443 Chantrey, Sir Francis Sketches for statue. May, 1834. Detail, side face,
full length, full face, seated – front, side
NPG Reference: 316A (27–30)

443.1 Chantrey, Sir Francis Pencil sketch – preliminary drawing for busts
and statue 1834 (inscribed and dated) 47.6 x 33.2 cms.
Given 1971
NPG Reference: 316A (29)

443.5 Ciba UK Ltd. Advertisement, *Financial Times*, 21 May 1970. 60 x 39
cms.
Text – A Quaker named John Dalton made an embarrassing mistake. He
went to church wearing red stockings
Illustration – Male model photographed wearing appropriate black clothing
and red stockings

444 Derome, J. Line and wash drawing. 12 x 9 cms.
Left profile holding measuring beaker. Almost identical with Stephenson
etching **460**
MAG Reference: 6560

445 Du Val, C.A. Etching of Chantrey statue. 1842. 18 x 12 cms.
Reproduced in *North of England Magazine*, 1842 (1), 1

446 Faulkner, B.R. Oil on canvas. 1840
Appears in the Royal Academy catalogue, 1840 but various authorities give
date as 1841 when it was presented by a memorial committee to the Royal
Society where it is now located. Reproduced *Times Educational Supplement*,
1956, 262

447 Faulkner, B.R. Oil on canvas. Undated
Copy of the Royal Society painting with slight difference in position of arms
and background. Presented, 1922, by Mr. David Dick, Glasgow
Manchester Town Hall

447.5 Gentleman, David (artist-designer) Postage stamp design [1965]
Based on the Allen portrait, with the words 'John Dalton, 1766–1844'
Reproduced, 5 x 6 cms., in an article by Richard Walter, 'How the GPO are
licking one problem' (*Sunday Times Magazine*, 10 July 1966, 42–43)
On 15 December 1964, the Postmaster General (Anthony Wedgwood Benn),
in a written House of Commons reply, stated that there would be an increase
in the number of special stamps 'to commemorate important anniversaries
and to reflect the British contribution to world affairs, including the arts and
science'. Accordingly, the Post Office was then reminded of the importance
of the approaching Dalton bicentenary in 1966 but, although the stamp was
designed by an eminent artist, the subject appears not to have been
considered appropriate enough to have come 'within licking distance of an
envelope'

448 Gilbert, Sir John and **Skill, J.F.** Eminent men of science living in the
years 1807–8 assembled in the library of the Royal Institution. Engraving by
William Walker. 1862. 51 x 83 cms.
The figure of Dalton is based on the Joseph Allen painting
NPG Reference: 1075, 1076A

449 Jeens, C.H. Engraving of right profile from daguerreotype [1896?] 22 x
14 cms.
Reproduced in Roscoe and Harden's *New view of the atomic theory* **490**
LP

450 Jordan, Charles Water-colour drawing [1840?].
Reproduced in E.M. Brockbank's *John Dalton*, 1944 **477**. Not available at time
of compilation.
Manchester Medical Society

451 Lonsdale, James Crayon. 1825. 51 x 41 cms.
Reproduced in Patterson **487.5**
MU Director of Finance (personal office)

452 Lonsdale, James Mezzotint. Artist: J. Lonsdale. Engraver: C. Turner:
Publisher: C. Turner. 1834. 25 x 20 cms.
'Original in the possession of James Thomson of Clitheroe' – Lonsdale's
Worthies of Cumberland, in which work it is reproduced
MPL, LP

453 Morton, W. Engraving of left profile. Undated. 18 x 11 cms.
MPL

454 Morton, W. Engraving of right profile. Undated. 7 x 6 cms.
MPL

455 Nicklin, John Daguerreotypes. 1842. 6 x 5 cms.
Probably three daguerreotypes were made at the same sitting and these are the only photographs of Dalton. A photographic enlargement of the one in the possession of the Society was made in 1961 and reproduced with the article by L.L. Ardern 'A unique John Dalton photograph', *Memoirs*, 1961–2, **104**, 67–9. One of the daguerreotypes (left profile) is in the Science Museum and the whereabouts of the third is unknown. See also **458**
LP, SM

455.5 Perigal, Arthur (?) Sketch full-face looking down, sepia tones. 1834 (?). 17 x 13 cms.
It seems likely that the sketch was made at the British Association meeting in Edinburgh, 1834. The portrait is discussed at some length in an article by E.C. Patterson (*Memoirs*, 1973–4, **116**, 12–5) where it is also reproduced
SNPG

456 Phillips, Thomas Oil on canvas. 1835. 92 x 71 cms. Signed at lower left with monogram 'TP' and dated 1835. Faintly inscribed 'Dr Dalton' at the left on a sheet showing chemical elements. (see p. 76)
Royal Academy catalogue, 1836, no. 415. Given by the artist to William Duckworth in November 1844 and described by W.C. Henry when in Duckworth's collection at Beechwood, Hants. Subsequently moved to his new house, Orchardleigh Park, Somerset, where it was sold by order of the late G.A.V. Duckworth's executors by Christie's, 21 September 1987
NPG

457 Pickard, Charles Colour print. Artist: Charles Pickard. Publisher: I.C.I. [1960?]. 23 x 19 cms.
LP

457.2 The Quaker Tapestry Tapestry, crewel embroidery. Scientists (panel D10). Panel designed and embroidered by Winifred Booker, 1984–5. 53 x 63 cms.
The Quaker scientists depicted are John Dalton, Arthur Eddington, and Kathleen Lonsdale. One of 77 panels. Reproduced as postcard and slide
Friends Meeting House, Kendal

457.5 Stephenson, James Pencil sketch for engraving. 15 x 12 cms. In frame 23.5 x 20 cms.
Left profile. Based on Nicklin daguerreotype **455**
Note on back of frame – 'Drawn by James Stephenson (1808–86), the engraver who was born in Manchester, apprenticed to John Fothergill of Prince's Court, Market Street, Manchester, and after his apprenticeship went

Portrait of Dalton by Thomas Phillips, 1835 **456**

to London, and entered the studio of William Finden. Among his contemporaries Stephenson was regarded as one of the finest line engravers in the country and, in vignette engraving, he was probably unsurpassed.' – Francis Nicholson, The Knoll, Windermere
MPL Reference: M628/3

458 Stephenson, James Engraving. Artist, engraver, publisher: James
Stephenson. 1842. 13 x 10 cms.
'The first page of a sheet of 4p. inviting subscriptions to the larger
engraving **459**. Page 3 is headed 'Meeting of the British Association…,
Manchester, June 1842'; and goes on 'the engraving will be executed by Mr.
James Stephenson and will be copied from an original drawing by himself,
greatly enlarged from a photographic portrait of the venerable doctor taken
during the present month'. See also **455**
MU

John Dalton by James Stephenson **460**

459 Stephenson, James Engraving. Artist, engraver and publisher: J. Stephenson. May, 1845. 29 x 25 cms.
Left profile. Based on the Nicklin daguerreotype.
MPL Reference: 7182

460 Stephenson, James Etching. Drawn, etched and published: J. Stephenson. 18 x 11 cms. (see p. 77)
Left profile holding measuring beaker
LP

461 XYZ (pseudonym) Scene, Royal Institution [Dalton thanking Gerrit Moll for defending English science]. Lithograph. Publisher: Crichton. 30 x 24 cms. See Cohen, Ernst 'Een physisch-chemische caricatuur' **377** Reproduced in Brockbank **477**, Paterson **487.5** and Thackray **492.5** CPL

SCENE ROYAL INSTITUTION,
Dedicated with permission to the British Association,
For the Advancement of Science. Printed at
A. Crichton's Lithographic Establishment 27 W. Registers.

Dalton (on the right) thanking Gerrit Moll for defending English science, *c.*1831 **461**

461.2 Artist unknown. Silhouette drawing. 25 x 15 cms.
MAG Reference: (1923.52)

462 Paper medallion. 1842. 2 x 2 cms.
Profile struck during the British Association Meeting, Manchester, 1842.
LP, NPG

Sculptures

463 Cardwell, Holme Marble bust. 1840
In Royal Academy Exhibition catalogue, 1840. Reproduced in E.M.
Brockbank's *John Dalton*, **477**. The original was in the Christie Library and is
now in Dalton-Ellis Hall. Casts are in LP and Cambridge University
Chemistry Laboratory.
MU

464 Chantrey, Sir Francis Bust. [1834?]
See the next two entries. Not traced

465 Chantrey, Sir Francis Statue (marble). 1837
This statue was a public tribute to Dalton resulting from the appeal made by
the Committee set up in 1833 **313**. It was brought to Manchester in 1838 and
exhibited at the Royal Manchester Institution from where it was transferred
to the Town Hall in 1882. With the statue of Joule, it shares the position of
honour at the entrance of the Town Hall. A newspaper clipping [c. 1888,
unidentified] in MPL gives a history of the statue.
'Dr. Henry says the likeness is more ideal than the reality, a refinement being
given to the countenance which did not exist in the bust which Chantrey first
took and used as a model when engaged on the full figure' – Smith's *Memoir*,
p. 275 **492**

466 Chantrey, Sir Francis Cheverton bust
'The intellectual bust, modelled by Sir Francis Chantrey, in preparation for
the statue and which has been reduced by a mechanical contrivance and
copied in ivory by Mr. Cheverton'. This seems to be the 'little bust'
presented by Dr. Adam Bealey who stated 'the bust is not ivory' He goes on
to say that 'we were assured by Mr. Clare, Dr. Dalton's executor, that the
mould was broken up when twelve were cast'. This may possibly refer to the
full-size bust. See *Proceedings*, 1910–11, **55**, 12–13, which includes a
photograph, plate 1. From a photograph of the interior of the Society's
premises the bust seems to have been about 15 cms. in height

467 Chantrey, Sir Francis? Circular plaque of head, left profile. 33 cms. in
diameter
Presented by Henry Galloway to Chetham's Library, 1886. Appears to be
based on the Chantrey bust
MC

468 Clarke, P.G. 'Sitting statue modelled from life (plaster), 1849. Purchased by the Committee' – entry in Peel Park, Salford, Museum catalogue, 1883 Probably a copy of the Chantrey statue **465**. Now destroyed

469 Levick, Ruby Bronze bust based on the Chantrey and Cheverton busts. 1903. 50 x 33 cms.
The bust was presented to the Society by Sir Henry Roscoe in 1903 and stood on the landing of 36, George Street alongside the special cabinet (also given by Roscoe) which housed the Dalton manuscripts. When the House was destroyed on 24 December 1940, the heavy bronze bust fell through the burning timbers into the deep cellars below and was buried in rubble. A successor building on the same site was opened in 1960 but became unsafe and in 1980 was sold for redevelopment. The contractors for the new office-block had to go deeper than the foundations of the 1960–80 premises and a mechanical excavator unearthed what was first thought to be a human head but eventually turned out to be a metal bust. The developers, French Kier Property Investments Ltd., had the bust expertly restored and, at a ceremony on 20 November 1981, presented it to the Society. See also **408, 465, 466, 809** and **831**
LP

469.2 Levick, Ruby Bronze bust. 1903.
Five coloured photographs of the bust, 20 x 15 cms., taken by the Yorkshire and Humberside Museum and Art Gallery Service, before restoration in 1981
LP

470 Smith, C.E. Statue (Caen stone). 1851.
Was situated at corner of Deansgate and John Dalton Street, Manchester. Now destroyed. See *Art Journal*, 1851, p. 61, and *Manchester Guardian*, 20 June, 1947

471 Smith, H.R. Profile in wax. 12 x 7 cms.
MU

472 Theed, William Statue (Bronze). 1854
A greatly enlarged copy of the Chantrey statue at the main entrance to the Town Hall erected as a public monument in Piccadilly, Manchester.
A plaster cast was presented to the Salford Art Gallery by W. Weekes, a London sculptor; now destroyed
In May 1966, the statue was moved to make way for a 33,000 volt primary substation of concrete, 20 feet underground. Its destination was the forecourt in Chester Street of the John Dalton College (now the Manchester Metropolitan University) which the Corporation thought 'a more appropriate setting'. 3 newspaper clippings in LP – *Guardian* 10 March 1966; *Guardian* un-dated; *Manchester Evening News*, 12 May 1966 (two photographs of the removal)
Manchester Metropolitan University

Film and television

472.2 ICI *The life and times of John Dalton, the founder of modern chemistry.* Film (black and white). 1966. Running time – 24 minutes.
Script – Michael Clarke; Producers – Gordon Begg, Michael Clarke; Director – Lawrence Moor
LP (MU)

472.4 BBC tv Tomorrow's World (programme 23), broadcast 7 April 1995
John Dalton's eyes 1995 Running time – 5 minutes
Presenter – Carol Vorderman. D.M. Hunt and J.D. Mollon describe how Dalton's genetic code was discovered and how the kind of colour blindness from which he suffered was established

472.6 BBC North West Home Ground, broadcast 7 February 1981
Manchester Literary and Philosophical Society Regional television programme 1981
Running time – 30 minutes
Presenter – Brian Redhead; Producer – Terry Wheeler; Director – Tony Wilkie-Miller
See also **997**

Illustrations

472.72 Marsh, J.T. Photographs of Pardshaw Hall Meeting House, Pardsey Crag, Eaglesfield Meeting House and Dalton's birthplace. 1967
7 photographs. 16 x 12 cms.
Useful annotations accompany each print
MPL

472.73 Dalton's birthplace, Eaglesfield. Exterior. c. 1880
Photograph. 16 x 24 cms. (See illustration on p. 82)
Originally in Dalton Hall. Reproduced in Greenaway **479**
SM; LP

472.74 Dalton's birthplace, Eaglesfield. Interior. c. 1880
Photograph. 16 x 24 cms. (See illustration on p. 83)
Originally in Dalton Hall. Reproduced in Greenaway **479**
SM; LP

472.75 Dalton's birthplace, Eaglesfield, from field at rear. n.d.
Pencil sketch. 15 x 19 cms.
Reproduced in Patterson **487.5**
CPL

473 Buss, R.W. British Association: visit of members to the fossil trees, 1842
Lithograph. Artist: R.W. Buss. Lithographer: G.H. Adcock. Publishers: Day

Exterior of Dalton's birthplace at Eaglesfield, Cumbria *c.*1880 **472.73**

and Haghe. 1843. 13 x 21 cms.

Although Dalton was probably too feeble to make this visit, he appears to have been the inspiration for one of the figures in the foreground.

Reproduced in J. Heywood's *Illustrations of the Manchester meeting of the British Association*

MPL

474 Buss, R.W. General meeting of the British Association for the Advancement of Science during the address of the President, Lord Francis Egerton, June 23, 1842

Lithograph. Artist: R.W. Buss. Lithographer: G.H. Adcock. [Publishers: Day and Haghe]. 1843. 13 x 21 cms.

Dalton is one of the members on the platform. Reproduced in J. Heywood's *Illustrations of the Manchester Meeting of the British Association*

MPL

474.5 Society's House – Secretary's room [1915?]

Photograph (Photographer not known). 21 x 16 cms.

Used as a laboratory by Dalton. See illustration on p. 127. Another

photograph, taken at a different angle, is included in Loudon **552**
LP

474.7 Shaw, James Society's House – Exterior. 1904
Photograph. 24 x 20 cms. The room used by Dalton is on the left of the
entrance. See illustration on p. 128.
LP

474.9 Manchester Literary and Philosophical Society Case of Dalton's
apparatus. [1915?]
Photograph. 20 x 15 cms.
Reproduced in Cardwell **478.2** plate IVa. The case is probably one of the
three mentioned by Barnes **379** of which two were in the library and one in
the passage of the old House; see illustration on p. 133.
LP

475 King, William John Dalton's house, Faulkner Street. 1857
Pencil and wash. 17 x 23 cms. (See illustration on p. 84)
MAG

475.2 Parry, James (?) St. Peter's Church and part of Mosley Street,
Manchester. (See illustration on p. 85)
Engraver: John Fothergill for T. Rogerson. Publisher: T. Sowler, 1824. 20.5 x
30 cms. Thought to be from a picture by James Parry, 1823.

Interior of Dalton's birthplace at Eaglesfield, Cumbria *c.*1880 **472.74**

Right-hand side: first building is the Assembly Rooms and next the
Independent Chapel. The present Mosley Street, from Booth Street to St.
Peter's Square, was originally called Dawson Street. Manchester New College
was on the corner of Dawson Street and Bond Street (now Princess Street),
probably the end building shown. Left-hand side: first building is the Portico
Library (still functioning) which Dalton usually visited twice a day to read the
newspapers. Next comes Charlotte Street which first crosses George Street
then Faulkner Street.

John Dalton's house, Faulkner Street, by William King, 1857 **475**

View of Mosley Street with the Portico Library on the left and St Peter's Church at the end, 1823 **475.2**

When Dalton first came to Manchester, he lived at the College in Dawson Street; he then had two successive addresses in Faulkner Street before joining the Johns family whose house was almost opposite the Lit and Phil premises in George Street. Four years after the Johns' moved in 1830, he set up house in Faulkner Street (facing Chain Street) where he lived until his death in 1844. During all his fifty years in Manchester, he never lived more than 300 yards away from the Portico and the Lit and Phil. Through this period, Mosley Street had changed from a quiet residential area into a busy thoroughfare full of warehouses and other businesses
MPL

475.4 Calvert, M. Cross Street Chapel. c. 1835
Pencil sketch
The oldest nonconformist congregation in Manchester, founded in 1662, had their chapel on this site from 1694. The building was damaged by Jacobites and rebuilt in 1715 and largely destroyed by enemy action in 1940. A new structure was completed in 1958 and demolished in 1995 to make way for an office development incorporating a chapel. A room at the rear of the building survived the bombing and this was the meeting place of the Society from 1781 to 1799. It was here that Dalton gave his first six lectures (**190–195**). Members of the chapel had an influence on the Town out of all proportion to their numbers (see McLachan. H. Cross Street Chapel in the

life of Manchester. *Memoirs*, 1939–41, **84** 29–41)
MPL

475.6 Harwood, J. The Friends Meeting House, Manchester. 1830
Engraving by R. Winkler. (See illustration on p. 131)
In January 1794, Dalton's certificate of membership of the Society of
Friends was received from Kendal by the Manchester Meeting. At this time
(1732), the Meeting House was at the corner of Jackson's Row and
Deansgate on the site of the original Quaker building of 1693. A new
Meeting House was erected at the corner of Dickenson Street and South
Street in 1795 to which Dalton subscribed 5gns. for the purchase of land.
The present building is on the same site extended to Mount Street and dates
from 1830; Dalton gave sums of £50 and £25 to this. It was here that the
1842 British Association meeting was held. Details are given in Marsh's 'Old
Quaker Dalton' **869**
MPL

475.8 Reston, Arthur Dog and Partridge Inn, Old Trafford, Manchester.
1884
Water colour. 9 x 14 cms.
Originally a farm, called Crowefeldeyate, with brewhouse, stables, gardens
and orchard, known to have been in existence in the sixteenth century. By
the late eighteenth century it had assumed its present name and was a
popular refreshment stop for the increasing number of coaches and wagons
on the busy main road to Chester (it is now set back from the present
alignment of Chester Road). The inn was considerably rebuilt in 1900.
Dalton had the habit of going to the Dog and Partridge 'every Thursday
afternoon, when the weather permitted, of taking exercise in the open air,
and of spending a few hours in company with a few intimate friends, in the
enjoyment of his favourite diversion of bowling'. See Henry **480** p. 209,
214–5

The Dog and Partridge Inn, Old Trafford, by John Owen, *c.* 1850 **475.8**

MPL Reference Rowbotham Collection
An earlier line drawing by John Owen c. 1850 20 x 26 cms.
MPL Reference Owen mss. vol. 40

475.9 Chetham's Library, Long Millgate, Manchester. 1950
Photograph. 20 x 16 cms.
Founded in 1653 by Humphrey Chetham, the Library is the oldest public
library in the English speaking world. In a letter to Elihu Robinson (20 Feb.

The reading room at Chetham's Library, Long Millgate, Manchester, *c.*1920 **475.9**

1794), Dalton writes 'There is in this town a large library, furnished with the
best books in every art, science and language, which is open to all gratis;
when thou art apprised of this…, thou wilt be able to form an opinion
whether I spend my time in slothful inactivity of body and mind.' Extracts
from an article by J. Bernouli in the *Transactions* of the Imperial Academy of
Science, St. Petersburg, 1727, copied in his own hand, still exist **106**, as does
the volume he actually used in the Library.
MPL Reference 100/28

476 Thomson, A.R. Dalton instructing two of his pupils.
Line drawing. 1955. 24 x 18 cms.
In Taylor, F.S. *An illustrated history of science* **741**

476.5 Dalton Plaque
Full size drawing of plaque by Manchester City Architect's Office. 1978. 55 x
55 cms.
Inscription: John Dalton (1766–1844) founder of the scientific atomic
theory President of Manchester Literary and Philosophical Society had his
laboratory here
LP

476.51 Correspondence relating to Dalton plaque
LP

476.52 Photographs of plaque
3 photographs. 14 x 14 cms.
1. Plaque. 2. Society's [1960–80] premises with plaque in position. 3. Plaque
refixed on successor building – Devonshire House, 36 George Street
LP

476.53 HRH the Duke of Edinburgh unveiling the plaque, 11 July 1979
Photograph. 16 x 12 cms.
LP

476.6 HRH the Duke of Edinburgh inspecting an exhibition of Dalton's
books, manuscripts and apparatus in the Dalton Room, Society's House, 36
George Street, 11 July 1979
6 photographs. 11 x 17 cms.
LP

476.7 Duckworth, E.H. Grave of John Dalton, Ardwick Cemetery,
Manchester. 18 February 1960.
2 photographs. 22 x 16.5 cms.
LP

Illustrations relating to Dalton

Apparatus **376, 478.2**; Atomic models **479, 761**; Atomic symbols **362.3,
362.4, 378, 479, 480, 686**; Birthplace **479, 540, 550, 577**; Funeral **363, 501,
502, 550**; Lecture sheets **378, 479, 686**; Notebook **490**; School **550**

?◆ Separately published works on Dalton

---

477 Brockbank, Edward Mansfield *John Dalton: some unpublished letters of
personal and scientific interest with additional information about his colour-vision and
atomic theories* Manchester, Manchester University Press, 1944. 62p. front.,

plates, ports., diagrs. 21 cms. (Half-title: Publications of the University of Manchester No. CCLXXXVII)

478 Brockbank, Edward Mansfield *John Dalton: experimental physiologist and would-be physician* Manchester, Falkner [1929]. 18p. ports., illus. 21 cms. (Cover-title: British Medical Association. Manchester and Salford meeting, July, 1929. History of medicine section)
Dalton's physiological papers. p. 18. A shortened version appears in *Brit. Med. J.,* 1929, (3589) 2, 730–2

478.2 Cardwell, D.S.L. (ed.) *John Dalton and the progress of science*: papers presented to a conference of historians of science held in Manchester, September 19–24 1966, to mark the bicentenary of Dalton's death. Manchester, Manchester University Press, 1968. xxii, 352p. ports., illus., 22 cms.
The following papers are included:
   Cardwell, D.S.L. Introduction
   John Dalton and the Manchester School of Science
   Hinshelwood, C.N. The qualitative and the quantitative
   Hall, M.B. The history of the concept of element
   Hall, A.R. Precursors of Dalton
   Guerlac, Henry The background of Dalton's atomic theory
   Thackray, Arnold Quantified chemistry – the Newtonian dream
   Kelham, B.B. Atomic speculation in the late eighteenth century
   Clow, Archie The industrial background to John Dalton
   Manley, Gordon Dalton's accomplishment in meteorology
   Farrar, K.R. Dalton's scientific apparatus
   Fox, Robert Dalton's calorific theory
   Greenaway, Frank Encounters with John Dalton
   Scott, E.L. Dalton and William Henry
   Brock, W.H. Dalton versus Prout: the problem of Prout's hypotheses
   Russell, C.A. Berzelius and the development of the atomic theory
   Crosland, M.P. The first reception of Dalton's atomic theory in France
   Farrar, W.V. Dalton and structural chemistry
   Solov'ev, Yu I. and Petrov, L.P. Russian scientists and Dalton's atomic
      theory
   Wright, W.D. The unsolved problem of 'daltonism'
   Urey, H.C. Dalton's influence on chemistry
These contributions are listed separately in the section beginning on page 112

478.5 Fleming, R.S. *John Dalton's development of a quantified chemistry: a reconstruction of the genesis of chemical atomism* 1981. PhD thesis, Cambridge University

479 Greenaway, Frank *The biographical approach to John Dalton* Manchester, Manchester Literary and Philosophical Society, 1958. Front. (port.), illus., facsim. [5], 98p (*Memoirs and proceedings*, 1958–9, **100**)
Includes illus. of exterior and interior of Dalton's birthplace; facsims. of mss. and lecture diagram; illus. of Dalton's atomic models

479.2 Greenaway, Frank *John Dalton and the atom* London, Heinemann, 1966. ix, 244p. plates, map, port., 22 cm. (Heinemann books on the history of science)

> Dear Miss Catharine Johns.
>
> The writer of this was born at the Village of Eaglesfield about 2 Miles west of Cockermouth, Cumberland. Attended the Village School there & in the neighbourhood till 11 Years of age, at which period he had gone through a course of Mensuration, Surveying, Navigation, &c; began about 12 to teach the Village School & continued 2 years afterwards; was occasionally employed in husbandry for a year or more; removed to Kendal at 15 Years of age as assistant in a boarding School, remained in that capacity for 3 or 4 years; then undertook the same School as a principal & continued it for 8 years, & whilst at Kendal employed his leisure in studying Latin, Greek, French & the Mathematics with Natural Philosophy: removed thence to Manchester in 1793, as Tutor in Mathematics & Natural Philosophy in the New College, was 6 years in that Engagement, & afterwards was employed as private & sometime public Instructor in various branches of Mathematics, Natural Philosophy & Chemistry, chiefly in Manchester, but occasionally by invitation in other places, namely London, Edinburgh, Glasgow, Birmingham & Leeds.
>
> Feb. 19th 1833   John Dalton

A letter from Dalton to Catharine Johns, 19 February 1833 **480**

480 Henry, William Charles *Memoirs of the life and scientific researches of John Dalton* London, printed for the Cavendish Society, 1854. xv, 249p. front. (port.), illus., 2 fold. pl., fold. facsim. 23 cms. (Half-title: Works of the Cavendish Society). Plates (1) facsim. autobiographical letter to Miss Catharine Johns, 19 February, 1833; (2) simple atmospheres; (3) atomic symbols, 1835
Appendix, p. 239–49, on colour-blindness by Dr. G. Wilson
MPL has Peel Memorial copy with 5p. added – 'Copy of a letter written by John Dalton to Elihu Robinson, of Eaglesfield, near Cockermouth, soon after Dalton had become a resident in Manchester, where, in 1793, he had been appointed tutor in mathematics and natural philosophy at the "Manchester Academy", now "Manchester College", Oxford. The letter contains probably the earliest account of that peculiarity of vision known as colour-blindness. The original has been presented to the Manchester Literary and Philosophical Society by Mr. Francis Nicholson, and is now in their possession'.

481 Kedrov, B.M. *The atomic theory of Dalton* [in Russian] Moscow, Academy of Sciences of the U.S.S.R. Institute of Philosophy; State Scientific and Technical Publishing House for Chemical Literature, 1949. 312p. front. (port.), illus., ports. diagrs.
MPL. mf.

481.7 Leitch, Diana and Williamson, Alfred *The Dalton tradition*; produced in conjunction with an exhibition to mark the 150th anniversary of the Royal Society of Chemistry, John Rylands Library, 16 April–26 June 1991. Manchester, John Rylands University Library of Manchester, 1991. 23p. illus., ports., facsims., 21 cms.
Includes a description of the conservation of the Dalton mss. in the Library's possession.

482 Lonsdale, Henry *The worthies of Cumberland. [vol. 5]. John Dalton* London, Routledge, 1874. xv, 320p. front. (port.), 1 fold. facsim.
Appendix 1, p. 301–8. John Dalton's statement of the affair betwixt his brother and self. Appendix 2, p. 309–20. List of Dr. Dalton's papers. The portrait is by J. Lonsdale

483 Meldrum, A.N. *Avogadro and Dalton: the standing in chemistry of their hypotheses* Edinburgh, W.F. Clay, 1904. 113p. 21cms.

484 Meldrum, A.N. *Avogadro and Dalton: the standing in chemistry of their hypotheses* Aberdeen, printed for the University, 1904. 113p. 28 cms. (Half-title: Aberdeen University studies, no. 10)

485 Meldrum, A.N. *The development of the atomic theory* London, Milford, Oxford University Press. [1920]. 13p. 23 cms.

Devoted very largely to vindicating the view that the atomic theory was not originated as a pure novelty by Dalton, but was a legitimate development of Newton's views. It is suggested that Dalton did not necessarily borrow his views from Higgins, but that both workers... followed much the same train of thought and reached essentially the same conclusions – *Nature*, 1920, **105** (2633) 212

486 Millington, John Price *John Dalton* London, J.M. Dent, 1906. xii, 225p. front. (port.), 19 cms. (Half-title: English men of science, ed. by J. Reynolds Green).
Reprinted AMS Press, 1971

487 Neville-Polley, Leonard Joseph *John Dalton* London, Society for Promoting Christian Knowledge, 1920. 63p. front. (port.), 19 cms. (Pioneers of Progress. Men of Science)

487.2 Nuffield Foundation, Science Teaching Project *Dalton and the atomic theory* [by **M.J.W. Rogers**] London, Longmans, Harmsworth Penguin, 1966. 10p. illus., port., facsim., diagr., 20 cms. (Chemistry background books, stage 2)

487.5 Patterson, E.C. *John Dalton and the atomic theory: the biography of a natural philosopher* New York, Doubleday, 1970. x, 348p illus., facsims., geneal., tabs., ports., 22 cm. (The science study series)
Although comparatively few new facts and original opinions are included, conveniently assembles much biographical information from other published sources. Well illustrated, with some useful notes and references at the end. Complements Thackray's biography **492.5**

488 Roscoe, Sir Henry Enfield *John Dalton and the rise of modern chemistry* London, Macmillan, 1895. 216p. front. (port.), illus., 2 fold. facsim, 19 cms. (The century science series)

489 Roscoe, Sir Henry Enfield *John Dalton and the rise of modern chemistry* London, Cassell, 1901. 216p. front. (port.), illus., 2 fold. facsims. 19 cms. (Half-title: The century science series, ed. by H.E. Roscoe)

490 Roscoe, Sir Henry Enfield and Harden, Arthur *A new view of the origin of Dalton's atomic theory: a contribution to chemical history; together with letters and documents concerning the life and labours of John Dalton, now for the first time published from manuscript in the possession of the Literary and Philosophical Society of Manchester* London, Macmillan, 1896. ix, 191p. front. (port.), 6 pl. (incl. facsims.). 23 cms
Contents: 1. On the genesis of Dalton's atomic theory. 2. Dalton's scientific diary, 1802–8. 3. Dalton's atomic weight numbers. 4. Notes of lectures 1810, 1814, 1818. 5. Letters written and received by Dalton. (Correspondents include – T.C. Hope, T. Thomson, W. Allen, H. Davy, J. Berzelius, J. Otley, L.

Sir Henry Roscoe (1833–1918) who with Arthur Harden was the first
biographer to make a detailed study of Dalton's scientific papers
488–491.2

Howard, P. Harris, D. Gilbert, D. Brewster, W. Whewell, H. Dalton).
Includes facsims. of 4 pages of notebook. The portrait is an engraving by
C.H. Jeens from the Nicklin daguerreotype

490.1 Roscoe, Sir Henry Enfield and Harden, Arthur A new view of the origin of Dalton's atomic theory. With a new introduction by Arnold Thackray. New York, Johnson Reprint, 1970. 225p. (Sources of Science, 100) Reprint of 1896 ed.

491 Roscoe, Sir Henry Enfield and Harden, Arthur *Die Entstehung der Dalton'schen Atomtheorie in neuer Beleuchtung. Ein Beitrag zur Geschichte der Chemie. Zugleich mit Briefen und Dokumenten über Dalton's Leben und Arbeiten, zum ersten Male aus dem im Besitze der Literary and Philosophical Society zu Manchester befindlichen Manuskripten veröffentlich* Ins Deutsche übertragen von Georg. W.A. Kahlbaum. Leipzig, J.A. Barth, 1898. xiv, 171p. front. (port.), 6pl. (incl. facsims.), tables, 23 cms.
(Added title-page: *Monographieen aus der Geschichte der Chemie*, hrsg. von Dr. Georg. W.A. Kahlbaum. 2 hft.)

491.2 Roscoe, Sir Henry Enfield and Harden, Arthur *Die Entstehung der Dalton'schen Atomtheorie in neuer Beleuchtung* Ins Deutsche übertragen von G.W.A. Kahlbaum. Leipzig, Zentralantiquariat der Deutsche Demokratischen Republik, [1970]
Reprint of 1898 ed.

492 Smith, Robert Angus *Memoir of John Dalton... and history of the atomic theory up to his time* London, H. Bailliere, 1856. v, [1], 298p. front. (port.), 22 cms.
(Also published with added title-page: *Memoirs of the Literary and Philosophical Society of Manchester*, 2nd ser., **vol. 13**). p. 253–63 list of Dalton's papers and publications. Portrait (front.) is Morton's engraving of Allen portrait with facsim. 'The weather is awkward; but the glass is rising. My best respects to all. Yours truly, John Dalton.'

492.2 Smyth, A.L. *John Dalton, 1766–1844: a bibliography of works by and about him* Manchester, Manchester University Press, 1966. xvi, 114p. ports., illus., 25 cm.
Lists 771 items relating to Dalton

492.5 Thackray, A.W. *John Dalton: critical assessments of his life and science* Cambridge, Harvard University Press, 1972. xiv, 190p. illus., 24 cm. (Harvard monographs in the history of science)
The most notable and perceptive contribution to modern Dalton studies, reflecting the author's systematic investigation of surviving Dalton mss. and other relevant original documents neglected by other writers. These new sources (many of which are reproduced in the text) are used in a critical analysis of the development of the chemical atomic theory. Brings to notice 'some of the graver misrepresentations in the received picture of Dalton'. The final chapter is a helpful 'bibliographical essay' on both manuscript and printed sources relating to Dalton

492.7 Thackray, A.W. *Introduction* to **Roscoe** and **Harden's** *A new view of the origin of Dalton's atomic theory* New York, Johnson Reprint, 1970. 25p.

492.9 USSR Academy of Sciences. Fundamental Library of the Social Sciences [Books and articles relating to John Dalton and selected works by him, 1802–1962]
Microfilm – 7 rolls
MPL Reference: mf540.1 Da2

?❧ Dalton's life: Periodical articles and book references 1830–1964

The main page reference only is given. Books are published in London unless otherwise stated.

1830–1899

493 Green, Jacob *Notes of a traveller, during a tour through England, France and Switzerland in 1828.* New York, 1830
Green's account of his meeting with Dalton in May, 1828 is reprinted in an article by E.F. Smith, 'Jacob Green, Chemist', 1790–1841. *J. Chem. Ed.*, 1943, **20**, 418–27

494 Wheeler, James, *Manchester: its political, social and commercial history, ancient and modern* 1836. p. 498–519
Probably the earliest published account of Dalton's life

495 Granville, A.B. *The spas of England* vol. 2, 1841. p. 17–8
Description of meeting with Dalton in his laboratory

496 Dr. Dalton (our portrait gallery) *North of England Magazine*, 1842, **1**, (1), 14–5
Includes an etching by C.A. Duval of the Chantrey statue

497 John Dalton [obituary] *Proc. Roy. Soc.*, 1843–50, **5**, 528–30

498 Dr. Dalton *Daisy Bank Magazine*, 1844, **1**, 112–5, 129–35, 145–8, 161–5

499 Pomps v. primitive simplicity – signs of the times *Gloucester Journal*, 7 Sept., 1844
A reply to this article, which commented on the alleged pomp of Dalton's funeral, appeared in the same paper dated 21? September. This was accompanied by a reprint of the letter in the *Manchester Guardian*, 14 Aug., 1844, from members of the Society of Friends giving their reasons for abstaining from attending the funeral. Protests against the manner of the

funeral also appeared in *The British Friend*, 31 Aug., 1844
Newspaper clippings in LP collection

500 Du Mênil, A.P.J. Daltons Tod. *Arch. Pharm., Berl*, 1844, **90**, (3), 321–5

501 The funeral of Dr. Dalton *Pictorial Times*, 24 August, 1844
Includes illustrations of funeral

502 Lying in state and funeral of the late Dr. Dalton *Illus. London News*, 1844, **5**, (120), 101–3
Includes illustrations of the lying in state and funeral procession

503 Thomson, Thomas Biographical account of the late John Dalton. *Proc. R. Phil. Soc. Glasg.*, 1844–8, **2** (2), 79–88

504 Biographic sketches: John Dalton *Chambers Edinburgh J.*, 1845, **3**, 211–5
Based on Wilson's article in the *British Quarterly Review*, 1845

505 John Dalton [obituary] *Annual Monitor of the Society of Friends*, 1845, N.S. 3, 40–7

506 John Dalton *Gentleman's Magazine*, 1845, N.S. 22, 431–2, 548–9

507 Wilson, George The life and discoveries of Dalton *British Quarterly Review*, 1845, 1, 157–98

508 Dalton *Fraser's Magazine*, 1854, 50, 554–72

509 Holland, Sir Henry Life of Dalton *Quarterly Review*, 1854–5, **96** (291), 43–53
Extended review of Henry's 'Life'

510 Trail, T.S. John Dalton *Encyclopædia Britannica*, 8th ed., vol. 7, 1854, p. 637–8

511 Eminent chemists: Dalton *Leisure Hour*, 1857, **6** (302), 650–3, 660–3

512 Review of Angus Smith's 'Life of Dalton' *Westminster Review*, 1857, N.S. 11, 270–4

513 Brewster, Sir David Memoirs of John Dalton *North British Review*, 1857, **27**, 465–97

514 Owen, Robert *The life of Robert Owen.* 1857, vol.1, p. 35–8

515 Holland, Sir Henry *Essays on scientific and other subjects* 1862. p. 386–42
Reprint of 'Life of Dalton' in *Quarterly Review*, 1854–5

516 Wilson, George Life and discoveries of Dalton. In *Religio chemici*. 1862. p. 304–64

517 Walker, William *Memoirs of distinguished men of science of Great Britain living in the years 1807–8.* 2nd ed. 1864. p. 41–4 See **448**

518 Gordon, M.M. *Home life of Sir David Brewster.* Edinburgh, 1869, p. 168–9 Describes Dalton at the British Association Meeting, 1842

519 Butler, F.H. John Dalton *Encyclopædia Britannica,* 9th ed., vol. 6, 1877, p. 784–7

520 Espinasse, Francis *Lancashire worthies.* 2nd ser. 1877 p. 261–73

521 Manchester City News *Notes and queries* Nancy Wilson (170), 30 Mar. 1878. Dalton's appearance (190), 6 Apr. 1878. 'Elements of English Grammar' (1262), 6 Sep. 1879

522 Stoughton, J. *Worthies of science.* 1879, p. 249–59

523 Slugg, J.T. *Reminiscences of Manchester fifty years ago.* Manchester, 1881, p. 106–8

524 Muir, M.M.P. *Heroes of science: chemists.* 1883, p. 106–54

525 Smith, R.A. *A centenary of science in Manchester.* 1883, p. 198–232. Also published as *Memoirs,* 1883, **29**

526 Clay, Charles A reminiscence of Dr. Dalton *Proceedings,* 1884, **23**, 83–7; *Chem. News,* 1884, **50**, 59–60

527 *Biographical catalogue of lives of Friends… whose portraits are in the London Friends Institute* 1888 p. 161–6

528 Reynolds, Osborne Memoir of James Prescott Joule. Manchester, 1892, p. 27–8, 80. *Memoirs,* 1892 **36**

529 T., W.A. Review of Roscoe's 'John Dalton and the rise of modern chemistry' *Nature* 1895, **52** (1338), 169–70

530 Reid, Sir Wemyss *Memoirs and correspondence of Lyon Playfair.* 1899 p. 57–8

1900–1929

531 Eaton, Seymour (ed.) *World's great scientists.* Chicago, 1900 p. 129–44. (Home study circle library, vol. 1)

532 Brittain, J.H. Manchester's contribution to the chemistry of the 19th century *Trans. Rochdale Lit. Sci. Soc.,* 1900–3, **7**, 4–9

533 Spencer, Herbert Distinguished dissenters. *Facts and comments,* 1902 p. 181–5

534 Stirling, William *Some apostles of physiology.* 1902, p. 86–7

535 Dalton celebrations in Manchester *Science,* 1903, N.S. **17**, 954–5

536 John Dalton: a sketch *Pop. Sci. Mon.,* 1903, **63**, 280–1

537 Manchester City News *Notes and queries* Lying in state (10353), 22 Oct. 1904; (10358), 29 Oct. 1904. Certificate (12762) [date not traced]

538 B., H. The forgotten: an interview with a statue: John Dalton and his work *Daily Dispatch,* 5 Jan., 1905

539 Hayes, L.M. *Reminiscences of Manchester… from the year 1840.* 1905. p. 5

540 Smalley, R. John Dalton's birthplace *Pharm. J.,* 1907, **79,** 841
Includes photograph of birthplace and words on memorial tablet

541 Roscoe, Sir H.E. Lecture on John Dalton delivered to the boys of Eton College In Roscoe, H.E., *Life and experiences.* 1906, Appendix 1, p. 395–404

542 Roscoe, Sir H.E. *Life and experiences.* 1906 p. 112–3

543 Clerke, A.M. John Dalton. *Dictionary of National Biography,* 1908. vol. 5, p. 428–34

544 Swindells, T. Manchester worthies: John Dalton's notable work for science *Manchester Evening Chronicle,* 5 Nov. 1910

545 Wenley, R.M. John Dalton and his achievement: a glimpse across a century *Pop. Sci. Mon.,* 1910, **76,** 500–12; *Sci. Amer,* 1910, splt. 70 (1805), 94–5

546 Nicholson, Francis Dr. Adam Bealey and Dr. Dalton *Proceedings,* 1910–11, **55,** 12–4
Recollections of one of Dalton's pupils. Dr. Bealey presented a small bust of Dalton to the Society. Photograph of bust (plate 1).

547 Roberts, Ethel *Famous chemists.* 1911, p. 63–71

548 Gibson, C.R. *Heroes of the scientific world.* 1912 p. 214–25

549 Griffiths, A.B. *Biographies of scientific men.* 1912, p. 114–25

550 Loewenfeld, Kurt Contributions to the history of science (period of Priestley, Lavoisier, Dalton) based on autograph documents *Memoirs,* 1912–13, **57** (19), 50p.
Includes illustrations of Dalton's funeral, his birthplace and the barn used by him as a school

551 Hird, F. *Lancashire stories* vol. 2, 1913, p. 425–31

552 Loudon, John Manchester memoirs. Article 10. *Royal Exchange Assurance Magazine,* 1916, **5,** 227–34
Includes photographs of the interior of the Society's House on the occasion of a special exhibition of Dalton relics at the time of the British Association meeting, 1915

553 Thomson, William Presidential address [on the Presidents of the Society] *Memoirs*, 1917–18, **62**, 1–14

554 Irwin, Wilfred An early chapter in the life of John Dalton *Friends Quarterly Examiner*, 1919, (211), 261–9

555 Harrison Frederic (ed.) *The new calendar of great men.* 1920. p. 443–5

556 Parkin, J.S. Greenrigg, Caldbeck *Trans. Cumberland and Westmorland Antiquarian Soc.*, 1921, **21**, 234–6
Greenrigg was for over 200 years the home of the maternal ancestors of Dalton

557 Tilden, Sir W.A. *Famous chemists, the men and their work.* 1921. p. 104–18

558 Manchester City News *Notes and queries* Theed statue (20115), 22 Aug 1922

559 Darrow, F.L. *Masters of science and invention.* 1923, p. 111–3

560 Nicholson, Francis The Literary and Philosophical Society, 1781–1851 *Memoirs*, 1923–4, **68**, 97–148
Includes various references to Dalton and his colleagues, also the text of a letter dated 20 Feb., 1794 to Elihu Robinson (p. 113–7)

561 Thorburn, A.D. Dalton memorials in Manchester, England *Industr. Engng. Chem.*, 1924, **16** (2), 190–1
Includes two illustrations of interior of the Society's old house. See also **919**

562 Hopwood, A. John Dalton *J. Chem. Ed.*, 1926, **3** (5), 485–91
Reprinted in Farber, E., *Great chemists*. 1961, p. 335–42

563 A Manchester beauty of 1793: Dalton's short love affair [1928?]
Unidentified newspaper clipping in Society's collection

564 Holmyard, E.J. *Great chemists.* 1928. p. 75–84 (The Great Scientists)

565 Bailey, R.W. The contribution of Manchester researchers to mechanical science *Proc. Instn. Mech. Engrs., Lond.*, 1929, **117**, 614–6
The article (p. 601–83) refers to a number of past members of the Society

566 Brindley, W.H. (ed.) *The soul of Manchester.* Manchester, 1929
Published for the Manchester section of the Society of Chemical Industry and edited by a former librarian of the Society, this volume includes a number of references to Dalton, particularly in the following chapters: A. Lapworth and J.E. Myers. Chemistry and Manchester University; C.L. Barnes. The Manchester Literary and Philosophical Society; John Allan. What chemistry has meant to industry in Manchester

567 Ostwald, Wilhelm Dalton. Bugge, G., *Das Buch der grossen Chemiker.* Berlin, 1929. vol. 1, p. 378–85, vol. 2, p. 488–9

1930–1964

568 Chance, Sir Frederick *Some notable Cumbrians.* Carlisle, 1931. p. 11–18

569 Holmyard, E. *Makers of chemistry* 1931. p. 221–40

570 Fry, A.R. *Quaker ways: an attempt to explain Quaker beliefs and practices and to illustrate them by the lives and activities of Friends of former days* 1933. p. 206–8

571 Kovanov, G. John Dalton *Khimiia i industria*, 1935, **13** (7), 298–300

572 Lenard, P. *Great men of science* 1933. p. 176–84

573 Paroushev, M. John Dalton (on the occasion of his 170th anniversary) *Khimiia i industria*, 1937, **15** (7), 357–9

574 Thomas, Henry and D.L. *Living biographies of great scientists* New York, 1941. p. 81–95

575 McLachlan, H. John Dalton and Manchester, 1793–1844 *Memoirs*, 1943–5, **86**, 165–75

576 Barnes, H.D. John Dalton: an address read at a joint meeting of the South African Chemical Institute and the University of Witwatersrand, August 3rd, 1944 *Iscor News*, 1944, 9

577 Friend, J. Newton John Dalton, 1766–1844 *Nature*, 1944, **154** (3899), 103–5
Illustration of Dalton's birthplace and Friends' Meeting House, Eaglesfield

578 Levy, J.F. The life of John Dalton *Bol. Soc. Quim, Peru*, 1944, **10**, 168–72

579 McKie, Douglas $H_2O + SO_3 = H_2SO_4$ *Manchester Evening News*, 27 July, 1944

580 Polanyi, Michael Reflections on John Dalton *Manchester Guardian*, 22 July, 1944

581 Price, H. John Dalton *Chem. & Drugg.*, 1944, **142**, 39; *Pharm. J.* 1944, **153**, 69

582 Thomas, John John Dalton as seen by Robert Owen *Manchester Evening News*, 11 Oct., 1944

583 W., J.R. John Dalton *Pharm. J.*, 1944, **153** (4215), 69

584 Memorial to John Dalton *Nature*, 1946, **158** (4006), 193
Plans for memorial stone in Pardshaw Hall graveyard, nr. Eaglesfield

585 Fleure, H.J. The Manchester Literary and Philosophical Society *Endeavour*, 1947, **6** (24), 147–51

586 Rives, H.E. and Forbush, G.E. *John book* New York, 1947. p. 216–7

587 Partington, J.R. John Dalton *Endeavour*, 1948, **7**, 54–6

588 Low, A.M. *They made your world* 1949. p. 69–72

589 Shippen, K.B. *Bright design* New York, 1949. p. 97–106

590 Smith, H.M. *Torchbearers of chemistry* New York, 1949. p. 59

591 McLachlan, H. John Dalton and Manchester, 1793–1844. *Essays and addresses.* 1950. p. 57–69

592 Holmyard, E.J. *British scientists* 1951. p. 37–9

593 Stevens, W.O. *Famous men of science* New York, 1952. p. 91–4 (Famous biographies for young people)

594 Elliott, T.L. John Dalton's grave *J. Chem. Ed.,* 1953, **30**, 569

595 Grigson, G. and Gibbs-Smith, C.H. (eds.) *People* 1954. p. 104–5

596 Ritchie, A.D. John Dalton: exponent of atomic chemistry *Times Educational Supplement* 1956, (2128), 262

597 Grogan, D.J. John Dalton, 1766–1844: a selection of books and papers by and about him *Manchr. Rev.,* 1957–8, **8**, 101–5

598 Hill, D.W. John Dalton: a toast, proposed by Dr. D.W. Hill on 25th October, 1957, [when] the Royal Institute of Chemistry held its 10th Dalton Lecture in Manchester *Manchr. Rev.,* 1957–8, **8** 97–101

599 Tanaka, Minoru Founders of chemistry: John Dalton *Kagaku,* 1958, **13**, 377–9

600 Cane, Philip *Giants of science* New York, 1959. p. 81–3

601 Irwin, K.G. *Men of chemistry* 1959. p. 47–54

602 Partington, J.R. Review of Greenaway's *The biographical approach to John Dalton. Nature,* 1959, **183** (4678), 1765

603 Brindley, W.H. Some notes on the history of the Society *Memoirs,* 1960–1, **103**, 60–9
Also published in *J. Inst. Chem.,* 1955, **79**, 62–9

604 Downs, R.B. *Molders of the modern mind* New York, 1961. p. 209–11

605 Knowles, Leo Peasant became genius of atom theory *Manchester Evening News,* 7 Sep. 1961

606 Carter, C.F. (ed.) Manchester and its region: a survey prepared for the [British Association] meeting held in Manchester, August 29 to Sep. 5, 1962. Manchester, 1962. p. 187–92
Plate 20, ms. pages of the *New System of Chemistry* [sic], vol. 2, pt. 1

607 Ardern, L.L. The Manchester Literary and Philosophical Society *J. Chem. Ed.*, 1962, **39**, 264–5
Includes reproduction of p. 1 of Paper 77

608 Treneer, Anne *The mercurial chemist: a life of Sir Humphry Davy.* 1963. p. 87–9

609 Shepherd, Walter *Great pioneers of science* 1964. p. 111–2

❧ Dalton's work: Periodical articles and book references 1800–1964

The main page reference only is given. Books are published in London unless otherwise stated.

1800–1849

610 Gough, John Strictures on Mr. Dalton's doctrine of mixed gases; an answer to Mr. Henry's defence of the same *J. Nat. Phil.*, 1804, **9**, 107–12

611 Harrington, Robert *The death-warrant of the French theory of chemistry, signed by truth, reason, common sense, honour and science… likewise, remarks upon Mr. Dalton's late theory…* 1804. p. 292–301
Harrington also wrote under the pseudonym Richard Bewley

612 Gough, John A reply to Dr. Dalton's objections to a late theory of mixed gases *Memoirs*, 1805, **6**, 405–23

613 Thomson, Thomas System of chemistry. 3rd ed. Edinburgh, 1807, vol. 3, p. 424–31, 451–2
The first publication of the principles of Dalton's atomic theory. 'This statement of Dalton's ideas is not quite exact' – Roscoe

614 Bostock, John Remarks on Mr. Dalton's hypothesis of the manner in which bodies combine with each other *J. Nat. Phil.*, 1811, **28**, 280–92
Critical of Dalton's atomic theory

615 Thomson, Thomas On the Daltonian theory of definite proportions in chemical combinations *Ann. Phil.*, 1813, **2**, 32–52, 167–71, 293–301; 1814, **3**, 11–8, 83–9

616 Higgins, William *Experiments and observations on the atomic theory and electrical phenomena* 1814
Copy in Manchester Central Library has list of errata crossed through followed by 'corrd. J.D.' in ms.

617 Macneven, W.J. Exposition of the atomic theory of chemistry and the doctrine of definite proportions *Ann. Phil.,* 1820, **16**, 195–214, 289–93, 338–50; *J. de Phys.,* 1821, **92**, 274–88, 376–401, 444–58

618 Benzenberg, J.F. *Über die Dalton'sche Theorie* Düsseldorf, 1830. xvi, 192p.

619 Daubeny, C.G.B. *An introduction to the atomic theory.* 1831. xv, 147p. Splt. 1840. xv, 62p. 2nd ed. Oxford, 1850. xxiii, 502p.
Letter from Dalton, p. 134–7

620 Thomson, Thomas The history of chemistry. 2nd ed. 1831, vol. 2, p. 285–308

621 Hume, G.L. *Chemical attraction: an essay in five chapters… 3. The atomic theory of Dr. Dalton* Cambridge, 1835

622 Whewell, William *History of the inductive sciences.* 1837. vol. 2, p. 501–23; vol. 3, p. 145–52

623 Davy, Sir Humphry *Collected works.* 1840. vol. 4, p. 78; vol. 5, p. 326–30; vol. 7, p. 93–9

624 Kopp, Hermann *Geschichte der Chemie* Brunswick, 1843. pt. 1, p. 362–8

1850–1899

625 Wilson, George On colour-blindness. Appendix p. 239–49, Henry's *Memoirs of Dalton,* 1854 **480**

626 Joule, J.P. Note on Dalton's determination of the expansion of air by heat *Memoirs,* 1858–60, **20**, 143–5

627 Lluch y Rafecas, F. *Teoría atómica… insiguiendo los principos de Dalton su fundador* S. Gervasio, Barcelona, 1862. 135p.

628 Scoffern, J. *Stray leaves of science and folk-lore* 1870. p. 43–63

629 Wurtz, A.C. *Geschichte der chemischen Theorie seit Lavoisier bis auf unsere Zeit* Berlin, 1870. p. 21–35

630 Kopp, Hermann *Die Entwicklung der Chemie* Munich, 1873. p. 285–300

631 Roscoe, Sir H.E. John Dalton and his atomic theory *Science lectures delivered in the Hulme Town Hall, Manchester.* 6th series, 1874, 15–27

632 Roscoe, Sir H.E. Some remarks on Dalton's first table of atomic weights *Proceedings,* 1875, **14**, 35–41; *Memoirs* 1876, **25**, 269–75; *Nature,* 1874–5, **11** (264), 52–4; *Chem. News,* 1878, **37**, 25–6

633 Masson, Herbert *Original research: the governing principles of the elements: explaining and supplementing Dr. Dalton's doctrine of definite reciprocal and multiple proportions...* 1878. v, 68p.

634 Wurtz, A.C. *The atomic theory*, trans. by A. Cleminson 1880. p. 23–32

635 Buckley, A.B. *A short history of natural science* 1883. p. 387–93

636 Muir, M.M.P. *A treatise on the principles of chemistry.* Cambridge, 1884. p. 8–12

637 Debus, Heinrich *Über einige Fundamentalsätze der Chemie, inbesondere das Dalton – Avogadro'sche Gesetz.* Cassel, 1894. viii, 99p.
Attempts to prove that Dalton was the real author of Avogadro's Law.

638 Marmery, J.V. *Progress of science.* 1895. p. 178–80

639 Roscoe, Sir, H.E. and Harden, Arthur A new view of the genesis of Dalton's atomic theory, derived from original manuscripts *Rep. Brit. Ass.,* 1895, **65**, 656

640 Debus, Heinrich Die Genesis von Daltons Atomtheorie *Z. Phys. Chem.,* 1896, **20** (3), 359–76; 1897, **24** (2), 325–52; 1899, **29** (2), 266–94; *Phil. Mag.,* 1896, **42**, 350–68 (part)

641 Wildermann, Meyer Experimenteller Beweis... des Daltonschen Gesetzes in sehr verdünnten Losüngen *Z. Phys. Chem.,* 1896, **19**, 233–50

642 Roscoe, Sir H.E., and Harden, Arthur The genesis of Dalton's atomic theory *Phil. Mag.,* 1897, **43**, 153–61

643 Roscoe, Sir H.E., and Harden, Arthur Die Genesis der Atomtheorie *Z. Phys. Chem.,* 1897, **22** (2), 241–9

644 Wildermann, Meyer Dalton's law in solutions *J. Chem. Soc.,* 1897, **71**, pt. 2, 743–55; *Z. Phys. Chem.* 1898, **25**, 711–21

645 Debus, Heinrich Erwiderung an Herrn Professor Kahlbaum *Z. Phys. Chem.,* 1899, **30** (3), 556–62

646 Kahlbaum, G.W.A. Bemerkung wider Herrn Heinrich Debus *Z. Phys. Chem.,* 1899, **29** (4), 700–4

1900–1929

647 Williams, H.S. *The story of nineteenth-century science.* 1900. p. 252–5

648 Ball, W.W.R. *A short account of the history of mathematics.* 1901. p. 441

649 Clarke, F.W. The atomic theory. The Wilde Lecture. *Memoirs,* 1902–3, **47** (11), 32p.

650 Divers, Edward The atomic theory without hypothesis *Rep. Brit. Ass.*, 1902, **72**, 557–75

651 Freund, Ida *The study of chemical composition.* Cambridge, 1904. p. 284–300

651.5 Roscoe, H.E. and Schorlemmer, C. A treatise on chemistry. 3rd ed. MacMillan, 1905. 2 vols.
Dalton's atomic theory vol. 1 p. 35–40

652 Meyer, Ernst von *A history of chemistry.* 1906. p. 196–204

653 Thomson, J.A. *Progress of science in the century.* 1906. p. 81–9

654 Arrhenius, Svante Theories of chemistry. 1907. p. 14–8, 39–46

655 Muir, M.M.P. *A history of chemical theories and laws.* 1907. p. 76–100

656 Larmor, J. On the physical aspect of the atomic theory. The Wilde Lecture. *Memoirs*, 1907–8, **52** (10), 54p.

657 Stange, Albert *Die Zeitalter der Chemie in Wort und Bild.* Leipzig, 1908. p. 254–60

658 Raykov P.N. Daltonovata teoriia. Neynoto istorichesko razvitié nastoyashté i badeshté i uspekhat na khimiiata pot neinoto rakovodstovo. (The Dalton theory: its historical development present and future, and the progress of chemistry under its influence) *Yearbook of the University of Sofia.* Official Part, 5, 1908–9, 37–60

659 Diergart, Paul *Beiträge aus der Geschichte der Chemie.* Leipzig, 1909, p. 530–52
Includes translation of Sir H.E. Roscoe's 'On the genesis of Dalton's atomic theory'

660 Meldrum, A.N. The development of the atomic theory
1. Berthollet's doctrine of variable proportions *Memoirs*, 1909–10 **54** (7), 16p.
2. The various accounts of the origin of Dalton's theory *Memoirs*, 1910–11, **55** (3), 12p. 3. Newton's theory and its influence in the eighteenth century *Memoirs*, 1910–11, **55** (4), 15p. 4. Dalton's physical atomic theory *Memoirs* 1910–11, **55** (5) 22p. 5. Dalton's chemical theory *Memoirs*, 1910–11, **55** (6), 18p. 6. The reception accorded to the theory advocated by Dalton *Memoirs*, 1910–11, **55** (19), 10p. 7. The rival claims of William Higgins and John Dalton *Memoirs*, 1910–11, **55** (22), 11p.

661 Thorpe, Sir Edward *History of chemistry.* 1909. vol. 1, p. 96–103

662 Brown, J.C. *A history of chemistry.* 1913. p. 317–23

663 Lowry, T.M. *Historical introduction to chemistry.* 1915. p. 292–305

664 Thorpe, Sir Edward Sir Henry Enfield Roscoe: a biographical sketch. 1916
p. 141–3 deal with the Roscoe-Debus controversy

665 Schuster, Arthur, and Shipley, A.E. *Britain's heritage of science.* 1917. p. 15–7

666 Scott, Alexander The atomic theory with especial reference to the work of Stas and Prout's hypothesis *J. Chem. Soc.,* 1917, **111**, 288–312

667 Libby, Walter *An introduction to the history of science.* 1918. p. 157–66

668 Sedgwick, W.T. and Tyler, H.W. *Short history of science.* New York, 1918. p. 361–3

669 Gregory, J.C. Dalton's debt to Democritus *Sci. Progr.,* 1919–20, **14**, 479–86

670 Delacre, Maurice *Histoire de la chimie.* Paris, 1920. p. 215–43

671 Delacre, Maurice Berzelius and Dalton *Monit. Sci.,* 1921, **11**, 1–8
Claims Berzelius erroneously coupled Richter's name with the law of definite proportions but from Dalton's 'law of the symbol' follow as corollaries the laws of definite and multiple proportions. Criticises Berzelius for not giving Dalton credit for introduction of atomic symbols. Summary in *Nature,* 1921, **107** (2692), 440

672 Meyer, Richard *Vorlesungen über die Geschichte der Chemie.* Leipzig, 1922. p. 76–80

673 Masson, Irvine, and Dolley, L.G.F. The pressures of gaseous mixtures *Proc. Roy. Soc.,* 1923, **103A**, 524–38
Discusses Dalton's law of additive pressure

674 Woodruff, L.L. (ed.) *Development of the sciences* New Haven, 1923. p. 84–6

675 Van Wagenen, T.F. *Beacon lights of science* New York, 1924. p. 211–3

676 Holmyard, E. *Chemistry to the time of Dalton* 1925. p. 111–25 (Chapters in the history of science)

677 Masson, Irvine *Three centuries of chemistry* 1925. p. 146–9

678 Coward, H.F. John Dalton: the early years of the atomic theory as illustrated by Dalton's own note-books and lecture diagrams: his apparatus *J. Chem. E.,* 1927, **4** (1), 23–37

679 Jakob, Max A simple proof of the failure of Dalton's law for actual gases *Z. Physik,* 1927, **41**, 737–8

680 Turner, D.M. *History of science teaching in England.* 1927. p. 54–5

681 Armitage, F.P. *A history of chemistry.* 1928 p. 63–8

682 Brockbank, E.M. John Dalton: experimental physiologist and would-be physician *Brit. Med. J.* 1929, **2** (3589), 730–2
Shortened version of **478**

1930–64

683 Ginzburg, Benjamin *The adventure of science.* New York, 1930, p. 200–18

684 Gregory, J.C. *A short history of atomism.* 1931. p. 65–94

685 Kirkby, William John Dalton, lecturer in pharmaceutical chemistry *Chem & Drugg.,* 1931, **115**, 16–7

686 Barclay, A. *Pure chemistry: a brief outline of its history and development.* 1937. pt. 1, p. 32–4; pt. 2, descriptive catalogue, p. 13–4
Includes illustrations of Dalton's lecture diagrams prepared in the Science Museum from the original drawings then held by the Society

687 Isaminger, M.P. Curious stories about health: discovery of color-blindness by John Dalton *Hygeia,* 1937, **15**, 207

688 Kedrov, B.M. Atomic theory of Dalton. [In Russian] *Pod Znamenem Marksizma,* 1937, **3**, 81–121
MPL mf.

689 Trattner, E.R. *Architects of ideas.* 1938, p. 72–97

690 Wolf, A. *A history of science, technology and philosophy in the eighteenth century.* 1938. p. 280–3
Dalton's meteorological writings

691 Ferchl, Fritz, and Suessenguth, A. A pictorial history of chemistry. 1939. p. 187–9

692 Moore, F.J. *A history of chemistry.* 3rd ed., 1939. p. 112–31

693 Partington, J.R. The origins of the atomic theory *Ann. Sci.,* 1939, **4** (3), 245–82

694 Pledge, H.T. *Science since 1500.* 1939. p. 117–8

695 Kedrov, B.M. *John Dalton, father of modern chemistry* [in Russian]. 1940. *In* 'Selected works on atomic theory' edited by the author. See **21**
MPL mf.

696 Evans, W.L. A present day examination of the postulates of John Dalton *Ohio J. Sci.,* 1941, **41**, 105–16

697 Singer, Charles *A short history of science.* 1941. p. 292–4

698 French, S.J. The drama of chemistry: 3, the chemical revolution *Dow Diamond*, 1942, **5** (3), 24–9

699 Rossiter, A.P. *The growth of science.* 1943. p. 94–7

700 Ritchie, A.D. The atomic theory *Memoirs*, 1943–5, **86**, 179–90
'My purpose is to try to show the part that Dalton played in converting an ancient theory… to scientific use…'

701 Carcamo, Victor Dalton and the atomic theory *Bol. Soc. Quim. Peru*, 1944, **10**, 173–81

702 Darwin, Sir Charles Atomic anniversary *The Listener*, 1944, **32** (825), 485–6

703 French, S.J. The drama of chemistry 1944. p. 34–7

704 Manley, Gordon John Dalton *Quart. J. R. Met. Soc.*, 1944, **70** (306), 235–9
Centenary tribute to Dalton's pioneer work in meteorology

705 Pereira Forjaz, D.A. O novo sistema de filosofia química de Dalton, 1766–1844 *J. Farmac.*, 1944, **3**, 5–9

706 Velde, A.J.J. van de John Dalton en de atoomtheorie *Verhandelingen van de Koninklijke Vlaamsche Academie voor Wettenschappen, Letteren en Schoon Kunsten van Belgie. Klasse der Wettenschappen*, 1944, **6** (11), 40p.

707 Musabekov, U.S. John Dalton [In the Azerbeijan language]. Four communications from the Azerbeijan branch of the U.S.S.R. Academy of Sciences, Baku. (3) 1945, 25–30 Résumé in Russian.
MPL mf.

708 Rousseau, Pierre Explorers of the atom from Dalton to Louis de Braglie and Joliot-Curie *La Nature*, **117**, 1945, 157–8

709 Historic researches: chemical elements and atoms *Engineer*, 1946, **182** (4728), 158–61

710 Berry, A.J. *Modern chemistry: some sketches of its historical development.* Cambridge, 1946. p. 1–3

711 Cheronis, N.D. *The study of the physical world.* 1946. p. 417–24

712 Butler, R.R. *Scientific discovery.* 1947. p. 168–70

713 Davies, Mansel *An outline of the development of science.* 1947. p. 136–8

714 Kedrov, E.M. The atomic theory of Dalton and its philosophical significance [In Russian]. *Izv. Akad. Nauk. SSSR*, 1947, **4** (6)
MPL mf.

715 Timmermans, J. *Histoire de la chimie.* Brussels, 1947. p. 35–9

716 Dampier, Sir W.C. *A history of science.* 4th ed. Cambridge, 1948. p. 208–13

717 Hooykaas, R. Dalton's atoom en molecuultheorie *Chem. Weekbl,* 1948, **44,** 229–37

718 Hooykaas, R. De Wordingsgeschiedenis van Dalton's theorie *Chem. Weekbl.,* 1948, **44,** 321–30

719 Hooykaas, R. Het Karakter van Dalton's theorie *Chem. Weekbl.,* 1948. **44,** 339–43

720 Hooykaas, R. De oorspronkelijkheid van Dalton's theorie *Chem. Weekbl.,* 1948, **44,** 407–11
Author's *Summary.* Dalton's conceptions are compared with older theories. Essentially his atomic theory is not a revival of democritean atomism but a continuation of an independent tradition of chemical corpuscular theories, adapted to newtonianism

721 Chalmers, T.W. *Historic researches: chapters in the history of physical and chemical discovery.* 1949. p. 115–24

722 Clusius, Klaus *Hundert Jahre Atomgewichtsforschung.* Munich, 1949. p. 5–9

723 Jaffe, Bernard Dalton: a Quaker builds the smallest of worlds. *Crucibles,* 3rd ed. 1949. p. 114–35 (4th ed. 1976)
A translation of the 1934 ed. into Russian, under the supervision of Prof. B.M. Berkenheim, appears in *Khimiya Shk.,* 1938, (3), 71–83
MPL mf.

724 Paneth, F.A. Review of Soddy's 'The story of atomic energy' *Nature,* 1950, **166** (4248), 799–800

725 Wightman, W.P.D. *The growth of scientific ideas.* 1950. p. 229–43

726 Partington, J.R. William Higgins and John Dalton *Nature,* 1951, **167** (4238), 120–1 and 1951, **167** (4253), 735–6
Disagrees with Soddy and Paneth who claim Higgins as the first to proclaim the modern atomic theory

727 Soddy, Frederick William Higgins and John Dalton *Nature,* 1951, **167** (4253), 734–5
Dispute with Partington over rival claims of Higgins and Dalton

728 Axon, H.J. The metallurgical writings of John Dalton *Bull. Inst. Metallurg.,* 1952, **3** (5), 13–7

729 Fierz-David, H.E. *Die Entwicklungsgeschichte der Chemie.* 2nd ed. Basel, 1952. p. 182–90

730 Leicester, H.M. and Klickstein, H.S. *A source book in chemistry.* 1952. p. 208–20

731 Melsen, A.G. van *From atomos to atom: the history of the concept atom.* Pittsburgh, 1952. p. 135–40

732 Wheeler, T.S. William Higgins *Endeavour,* 1952, **11** (41), 47–52

733 Crerar, J.W., and Ross, J.A. John Dalton... Captain Joseph Huddart,... and the Harris family. Historical notes on congenital colour-blindness *Brit. J. Opthal.,* 1953, **37**, 181–4

734 Greenaway, Frank John Dalton and the rebirth of atomism *Discovery,* 1953, **14**, 318–20

735 Papp, Desiderio Cuál es el origen gnoselológico de la teoria atomica de Dalton? *Arch. Int. Hist. Sci.,* 1953, **6** (23–4), 232–48
Communicated to the 6th International Congress of the History of the Sciences, Amsterdam, 1950

736 Duveen, D.I., and Klickstein, H.S. John Dalton's autopsy *J. Hist. Med.,* 1954, **9**, 360–2
Mainly on Dalton's brain and eyes

737 Kedrov, B.M. Jubilee of the atomic theory and the Russian chemists [In Russian]. *Vestnik Leningradskovo Universiteta,* 1954, (5), 177–84
From materials in the Mendeléeff Museum Archives, Leningrad University, includes letters of Mendeléeff, Tischchenko, Beketov and Boyd Dawkins in connection with the celebration of the Dalton Centenary, 1903. The letters comprise an evaluation of the discoveries of Dalton
MPL mf.

738 Major, R.H. *History of medicine.* 2nd ed. 1954. p. 810

739 Partington, J.R. Seventeenth-century chemistry, the phlogiston theory and Dalton's atomic theory *Nature,* 1954, **174** (4424), 291–3

740 Partington, J.R. Dalton's atomic theory *Scientia,* 1955, **90** (7), 221–5

741 Taylor, F.S. *An illustrated history of science.* 1955. p. 105–8
Includes illustration by A.R. Thomson of Dalton instructing two of his pupils

742 Leicester, H.M. *The historical background of chemistry.* 1956. p. 153–8

743 Nash, L.K. The origins of Dalton's chemical atomic theory *Isis,* 1956, **47** (148), 101–16

744 Erdélyi, Stefan Das Problem von Dalton und die Entropie *Wärmetechnik,* 1957, **8**, 15–6

745 Nash, L.K. (ed.) Atomic-molecular theory
In Conant, J.B. (ed.), *Harvard case histories in experimental science*, Cambridge, Mass., 1957. vol. 1, p. 215–321

746 Peugler, H. John Dalton: der Schöpfer der Atomwissenschaft *Ausbau*, 1957, **10**, 517–25

747 Read, John *Through alchemy to chemistry: a procession of ideas and personalities.* 1957. p. 147–57

748 Green, J.H.S. Two classics of chemistry *Sci. Progr.*, 1958, **46**, 429–40
Review of the revolutionary contributions to chemical theory made by Dalton's *New System of Chemical Philosophy*

749 Schwartz, G., and Bishop, P.W. *Moments of discovery.* Vol. 2. Development of modern science. 1958. p. 768–70

750 Hull, L.W.H. *History and philosophy of science.* 1959. p. 265–9

751 Reason, H.A. *The road to modern science.* 3rd ed. 1959. p. 101–4

752 Singer, Charles *A short history of scientific ideas to 1900.* Oxford, 1959. p. 343–4

753 Gillispie, C.C. *The edge of objectivity.* 1960. p. 252–9

754 McKenzie, A.E.E. *The major achievements of science.* Cambridge, 1960. vol. 1, p. 133–44

755 Partington, J.R. *A short history of chemistry.* 3rd ed. 1960. p. 167–79

756 Szmid, J. Dalton und seine Theorie *Chemie in der Schule*, 1960, **7**, 85–96

757 Turner, D.M. *The book of scientific discovery.* 3rd ed. 1960. p. 140–3

758 Wheeler, T.S., and Partington, J.R. *The life and work of William Higgins, chemist, 1763–1825.* 1960. p. 123–42

759 Dalton begründet den Begriff des Atomgewichts *Wissenschaft und Fortschritt*, 1961, **11**, 251

760 Anthony, H.D. *Science and its background.* 4th ed. 1961. p. 215–9

761 Greenaway, Frank Chemistry at the Science Museum *J. Inst. Chem.*, 1961, **85**, 126–31
Includes illustrations of Dalton's atomic models

762 Guerlac, Henry Some Daltonian doubts *Isis*, 1961, **52**, 544–54
Suggests that Dalton was influenced by Richter's ideas

763 Crosland, M.P. *Historical studies in the language of chemistry.* 1962. p. 256–64
Discusses Dalton's atomic symbols

764 Klyachko, L.S., and Zhorovovich, T.P. John Dalton and his works [In Russian]. *Inzh.-fiz. Zh.* 1962, **5** (3), 132–3
MPL mf.

765 Morgan, Bryan *Men and discoveries in chemistry.* 1962. p. 72–84

766 Partington, J.R. *A history of chemistry.* 1962. vol. 3, p. 749–822
One of the most detailed modern accounts of Dalton's work, with numerous bibliographical references. Higgins's claim to have anticipated Dalton is dealt with on p. 749–54

767 Toulmin, Stephen and Goodfield, June *The architecture of matter.* 1962. p. 228–36

768 Forbes, R.J. and Dijksterhuis, E.J. *A history of science and technology.* 1963. vol. 1, p. 225–9

769 Garrett, A.B. *The flash of genius.* 1963. p. 68–72, 162–9

770 Moulton, F.R., and Schifferes, J.J. (eds.) *The autobiography of science.* 1963. p. 231–4

771 Wightman, W.P.D. Science in the eighteenth and nineteenth centuries In Brierley, John (ed.), *Science in its context,* 1964. p. 187–8

772–799 not used

❧ Articles in periodicals and composite works, 1965–95

800 Axon, Howard John Dalton shows his metal *Memoirs,* 1990–1, **130,** 120–2
Notes the early nineteenth century demand for information on metallic alloys, also the nomenclature 'Daltonide' and 'Bertholide' for intermetallic compounds. Concludes that in relation to metal alloy systems, Dalton's basic ideas were undoubtedly productive and in advance of his time, but their application was faulty

801 Barr, E.S. Anniversaries in 1966 of interest to physicists *Am. J. Phys.,* 1966, **34,** 31–5

802 Bedlow, Robert Atom pioneer's charred papers saved *Daily Telegraph,* 16 April 1991, p.6

803 Bernatowicz, A.J. Dalton's rule of simplicity *J. Chem. Ed.,* 1970, **47,** 577–9

804 Bolam, C.G. John Dalton, 1766–1844 *The Inquirer*, 1966, Sept., p.3

805 British Biographical Index *British biographical archives, 1990: John Dalton*, fiche 300, p. 212–49
Extracts from 13 (mainly 19th century) biographical reference books

806 Brock, W.H. Dalton versus Prout: the problem of Prout's hypotheses (In Cardwell, D.S.L. *John Dalton and the progress of science*, p.240–58) **478.2**

807 Brock, W.H. John Dalton: natural philosopher or chemist? *Education in Chemistry*, 1994, **31** (4), 95–6, 102

808 Butterworth, L.M.A. Dr John Dalton, 1766–1844 *Friends Q.*, 1977, **20**, 135–8

809 Cardwell, D.S.L. and Mottram, Joan Fresh light on John Dalton *Notes and Records Roy. Soc.*, 1984, **39** (1), 29–40
Reports on the recent disinterment of the Levick bust of Dalton **469**; account of Dalton written by John Roberton, 13 September 1844

810 Cardwell, D.S.L. John Dalton and the Manchester school of science (In Cardwell, D.S.L. *John Dalton and the progress of science*, p.1–9) **478.2**

811 Cardwell, D.S.L. Two centuries of the Manchester Lit and Phil *Memoirs*, 1980–1, **121**, 122–37
A lecture delivered at Cross Street Chapel on the occasion of the two hundredth anniversary of the first meeting of the Society almost to the hour, that is at about 7.30 in the evening of March 15th, 1781. It is also the best modern account of the Society's history

812 Clow, Archie The industrial background to John Dalton (In Cardwell, D.S.L. *John Dalton and the progress of science*, p.125–39) **478.2**

813 Cole, Theron Dalton, mixed gases, and the origin of the chemical atomic theory *Ambix*, 1978, **25** (2), 117–30

814 Crosland, M.P. The first reception of Dalton's atomic theory in France (In Cardwell, D.S.L. *John Dalton and the progress of science*, p. 274–89) **478.2**

815 Da Costa, A.M.A. Simplicity and verisimilitude in chemistry: John Dalton and Karl Popper [in Portuguese] *Bol. Soc. Port. Quim.*, 1987, **27**, 13–8

816 Dalton, John Bicentenary of the birth of John Dalton *Engineer*, 1966, **222**, 756

817 Dalton, John John Dalton, letters to his brother: with brief notes by A.L. Smyth *Memoirs*, 1993–4, **132**, 160–76
Letters dated 24 May 1792, 6 March 1799, 6 February 1805 and 30 October 1817

818 Ditchfield, G.M. The early history of Manchester College *Trans. Historic Soc. of Lancashire and Cheshire*, 1972, **123**, 81–104
Covers the history of the College to 1803

819 Dowling, Rob The law of multiple proportions *Chem. 13 News*, 1979, **104**, 3
Describes a laboratory experiment illustrating the law

820 Elliott, R.L. John Dalton: the facts concerning his colour blindness *J. Soc. Dy. Col.*, 1984, **100**, 319–21

821 Elsen, K.M. The law of multiple proportions *Chem. 13 News*, 1980, **110**, 12–3
An experiment in the law of multiple proportions is demonstrated by the decomposition of a compound containing only copper and bromine to a second compound containing the same elements

822 Emery, A.E. John Dalton (1766–1844) *J. Med. Genet*, 1988, **25**, 422–6

823 Farrar, K.R. Dalton's scientific apparatus (In Cardwell, D.S.L. *John Dalton and the progress of science*, p. 159–86) **478.2**
Well illustrated extensive description of Dalton's apparatus. Rejects Roscoe's 'ink bottle legend'

824 Farrar, W.V. Dalton and structural chemistry (In Cardwell, D.S.L. *John Dalton and the progress of science*, p. 290–9) **478.2**

825 Farrar, W.V. Nineteenth century speculations on the complexity of the chemical elements *Brit. J. Hist. Sci.*, 1965, **2**, 247–323

826 Farrar, W.V. and K.R. and Scott, E.L. The Henrys of Manchester, pt. 3 William Henry and John Dalton *Ambix*, 1974, **21**, (2–3), 208–28
Describes the friendship between Henry and Dalton, especially the atomic theory of Dalton based on work done by Henry

827 Farrar, W.V. and K.R. and Scott, E.L. Y teulu Henry ym Manceinion [in Welsh] *Y Gwyddonydd*, 1971, **9**, 57–63

828 Feigl, D.M. Dalton's law experiment for students in the health sciences *J. Chem. Ed.*, 1974, **51** (4), 273–4
Measurement of carbon-dioxide in inhaled and exhaled air as an illustration of Dalton's law of partial pressures

829 Fleming, R.S. Newton, gases, and Daltonian chemistry: the foundations of combination in definite proportion *Ann. Sci.*, 1974, **31** (6), 561–74

830 Fox, Robert Dalton's caloric theory (In Cardwell, D.S.L. *John Dalton and the progress of science*, p. 187–202) **478.2**

831 French Kier Holdings The development company's contribution to scientific history *FK News*, 1982, (10) Summer
Describes the unearthing of the Levick bust of John Dalton **469**. Colour photographs of the bust and the presentation of the bust to the President of the Lit. and Phil. (Mr David Wilson) by the Group Chairman (Mr J.C.S. Mott). The Company paid for the restoration of the bust

832 Fujii, Kiyohisa The Berthollet-Proust controversy and Dalton's chemical atomic theory *Brit. J. Hist. Sci.*, 1986, **19**, 177–200

833 Goodman, D.C. Wollaston and the atomic theory of Dalton *Hist. Stud. Phys. Sci.*, 1966, **3** (9), 1–23

834 Greenaway, Frank Dalton and Joule: a bridge between two worlds *Memoirs*, 1989–90, **129**, 17–28
The opening address of the Joule Centenary Celebration 1989, which describes how Dalton and his pupil bridge the years between 18th century philosophy and 20th century science

835 Greenaway, Frank The Dalton bicentenary celebration *Chem. Brit.*, 1966, **2** (12), 546–7

836 Greenaway, Frank Encounters with John Dalton (In Cardwell, D.S.L. *John Dalton and the progress of science*, p. 203–19) **478.2**

837 Greenaway, Frank John Dalton *Endeavour*, 1966, **25**, 73–8

838 Greenaway, Frank John Dalton as a historical figure *Nature*, 1966, **211**, 1013–4

839 Greenaway, Frank John Dalton in London *Proc. Roy. Instn.*, 1966, **41** (189), 162–77

840 Greenaway, Frank A man of atomic proportions *Chem. Brit.*, 1994, **30** (11), 920–1

841 The Guardian John Dalton [supplement] *The Guardian* 19 September 1966, p. 6–7
Authors of articles include J.T. Marsh and Gordon Manley; appropriate supporting advertisements

842 Guerlac, Henry The background to Dalton's atomic theory (In Cardwell, D.S.L. *John Dalton and the progress of science*, p. 57–91 **478.2** and the author's *Essays and papers in the history of modern science*, p. 217–44)

843 Hall, A.R. Precursors of Dalton (In Cardwell, D.S.L. *John Dalton and the progress of science*, p. 40–56) **478.2**

844 Hall, Christopher John Dalton and the discovery of colour blindness *Ophthalmic Opt.*, 1968, **8**, 920–2

845 Hall, M.B. The background of Dalton's atomic theory *Chem. Brit.*, 1966, **2**, 341–5

846 Hall, M.B. The history of the concept of element (In Cardwell, D.S.L. *John Dalton and the progress of science*, p. 21–39) **478.2**

847 Hartley, Sir Harold Dalton as genetic sport *New Scientist*, 1966, **32**, 255 Review of Frank Greenaway's *John Dalton and the atom*

848 Hartley, Sir Harold John Dalton, FRS (1766–1844) and the atomic theory: a lecture to commemorate his bicentenary *Proc. Roy. Soc. A.*, 1967, **300**, 291–315

849 Hatton, A.P. and Flowett, J.W. Clegg's model of a Watt beam engine *Memoirs*, 1963–64, **106**, 104–7
Samuel Clegg, the gas engineer (*DNB*), was a pupil of Dalton's and served his apprenticeship with Boulton and Watt. About 1806 he presented a working model of a double-acting single cylinder beam rotative engine to Dalton who used it in his lectures. The model survived the destruction of the Society's building and its renovation is described. It is now displayed in the Manchester Museum of Science and Industry **1008**

850 Higasi, Keniti Dalton's atomic theory and his law of partial pressures: a case of breakthrough in chemical research *Kagaku Kyoiku*, 1976, **24** (4), 327–31

851 Hijioka, Yoshito and others Review of studies of John Dalton, 1766–1844 [in Japanese] *Kagakushi*, 1978, **7**, 46–53

852 Hinshelwood, Sir Cyril The qualitative and the quantitative *Memoirs*, 1966–7, **109**, 18–26
(Also in Cardwell, D.S.L. *John Dalton and the progress of science*, p. 10–20) **478.2**

853 Hunt, D.M., Dulai, K.S., Bowmaker, J.K., Mollon, J.D. The chemistry of John Dalton's colour blindness. *Science*, 1995, **267** (5200), 984–8
DNA extracted from Dalton's preserved eye tissue showed that Dalton was a deuteranope, lacking the middlewave photopigment of the retina. The diagnosis is compatible with the historical record of his phenotype, although it contradicts Thomas Young's belief that Dalton was a protanope.

854 Iyama, Hiroyuki A case of fabricated discovery: the law of multiple proportions *Hist. Scientiarum*, 1983, **24**, 19–28

855 not used

856 Iyama, Hiroyuki John Dalton and his law of multiple proportions (in Japanese) *Kagakushi*, 1981, **17**, 24–32

857 Iyama, Hiroyuki The role of John Dalton in the history of the physical sciences (in Japanese) *Kagakushi Kenkyu*, 1982, **21**, 129–41

858 Jensen, A.T. John Dalton, 1766–1966 *Dansk Kemi*, 1966, **47** (11), 173–4

859 Jones, D.E.H. The atomization of chemistry *New Scientist*, 1966, **31**, 493–6

860 Kelham, B.B. Atomic speculation in the late 18th century (In Cardwell, D.S.L. *John Dalton and the progress of science*, p. 109–24) **478.2**

861 Kleinman, G.S. Dalton bicentennial: a teacher by choice *Science Education*, 1966, **50** (5), 464–6

862 Klyachko, L.S. Unity of Dalton and Stefan solutions to the problem of liquid evaporation from an open surface *Zh. Fiz. Khim.*, 1974, **48** (5), 1123–5
In the evaporation of solutions, the laws of Dalton (empirical) and Stefan (theoretical) are both applicable

863 Knight, D.M. The atomic theory and the elements *Stud. Romant.*, 1966, **5**, 185–207
Deals especially with Dalton and Davy in the period 1800–1820

864 Krasovitskaia, T.I. and Plotkin, S.Ia. Vollaston i atomnia teoria Dal'tona [in Russian] (Wollaston and Dalton's atomic theory) *Vop. Ist. Est. Tekh.*, 1973, **45**, 41–4

865 Laszlo, Vörös Dalton, 1766–1844 [in Hungarian] *Termeszetterdomanyi Közlöny*, 1966, **10** (7), 289–91

866 Manley, Gordon Dalton's accomplishment in meteorology (In Cardwell, D.S.L. *John Dalton and the progress of science* p. 140–58) **478.2**

867 Manley, Gordon John Dalton's snowdrift *Weather*, 1952, **7**, 210–2

868 Marsh, J.T. John Dalton as a Manchester Friend *The Friend*, 1966, **124** (37), 1087–9

869 Marsh, J.T. Old Quaker Dalton *Memoirs*, 1968–9, **111**, 27–47 Also in *Friends Quarterly*, 1970, **16** (10), 519–36

870 Mauskopf, S.H. Haüy's model of chemical equivalence: Daltonian doubts exhumed *Ambix*, 1970, **17**, 182–91

871 Mauskopf, S.H. Thomson before Dalton: Thomas Thomson's considerations of the issue of combining weight proportions prior to his acceptance of Dalton's chemical atomic theory *Ann. Sci.*, 1969, **25**, 229–42

872 Mendoza, Eric A critical examination of Herapath's dynamical theory of gases *Brit. J. Hist. Sci.*, 1975, **8** (2), 155–65

873 Meredith, W.J. 'What Manchester thinks…' *Brit. J. Radiology*, 1968, **41**, 2–11
Starting with Dalton, traces the Manchester contribution to radiology

874 Miles, W.D. John Dalton's autopsy *J. Hist. Med.*, 1957, **12**, 263–4

875 Mollon, J.D., Dulai, K.S. and Hunt, D.M. Dalton's colour blindness: an essay in molecular biography
(In Carden, D. and others (eds.) *John Dalton's colour vision legacy*) **993**

876 Mortimer, C.E. John Dalton: color blind chemist
in *Muhlenberg essays in honor of the college centennial*, Allentown, Penn.
Muhlenberg College, 1968, p. 271–86

877 Murgulescu, I.G. John Dalton [in Romanian] *Stud. Cercet. Chim.*, 1966, **14**, 587–92

878 Nicholas, J.W. The Dalton anniversary *Cumbria*, Sept. 1966, 294–5
Includes information on Dalton's birthplace

879 Noble, Vernon The quiet countryman *Lancashire Life*, 14 Sept. 1966, 70–5
Includes photographs of Dalton's cradle, barometer and spectacles, as well as the Clegg engine used in lectures

880 Patterson, E.C. John Dalton and the atomic theory [reviews] **487.5** *J. Amer. Chem. Soc.*, 1971, **93**, 2577 *Physics Today*, 1972, **25**, 68

881 Patterson, E.C. John Dalton in Edinburgh *Memoirs*, 1973–4, **116**, 5–19
Dalton's most important visit to Edinburgh was in 1807 when, as a comparatively little known schoolmaster, he gave a short series of subscription lectures at the invitation of Thomas Thomson. This paper is mainly concerned with the British Association Meeting of 1834 where he was honoured as the internationally renowned scientist. He gave no papers of his own but took part in the discussion

882 Ponte, P.L. The contribution of John Dalton to medicine chemistry and physics in the second century after his birth *Minerva Med.*, 1967, **58** (12), 200–4
Emphasizes Dalton's contribution to instrumentation, colour vision, gas mixtures, law of multiple proportions, and atomic and molecular chemistry

883 Ramage, Donald John Dalton: bicentenary of a great man of science *The Friend*, 2 Sept. 1966, **124** (35), 1021–3

884 Raymond, Jean and Pickstone, J.V. The natural sciences and the learning of the English Unitarians: an explanation of the roles of the Manchester College
(In Smith, Barbara. *Truth, liberty, religion*, p. 127–64)

885 Rocke, A.J. Atoms and equivalents: the early development of the chemical atomic theory *Hist. Stud. Phys. Sci.*, 1978, **9**, 225–63

886 Rocke, A.J. The reception of chemical atomism in Germany *Isis,* 1979, 70 (254), 519–36

887 Rodda, J.C. Eighteenth century evaporation experiments *Weather,* 1963, **18**, 264–9
Observations by Dobson (1777) and Dalton (1802)

888 Rose, J. John Dalton the chemist *Nature,* 1966, **211**, 1015–6

888.5 Rouvray, D.H. John Dalton: the world's first stereochemist *Endeavour,* 1995, N.S. **19** (2) 52–7

889 Russell, C.A. Berzelius and the development of the atomic theory
(In Cardwell, D.S.L. *John Dalton and the progress of science,* p. 259–73) **478.2**

890 Scott, E.L. Dalton and William Henry
(In Cardwell, D.S.L. *John Dalton and the progress of science,* p. 220–39) **478.2**

891 Sheng, Genyu John Dalton: father of chemistry of modern times [in Chinese] *Ziran Zazhi,* 1983, **6**, 863–8

892 Siegfried, Robert Further Daltonian doubts *Isis,* 1963, **54**, 400–81

893 Simms, Colin Note on the herbaria of John and James Dalton *J. Soc. Biblphy. Nat. Hist.,* 1969, **5** (2), 117–20
Reports confusion between John Dalton and the Revd James Dalton (1764–1843), a noted Yorkshire botanist See also **104, 104,1** and **380**

894 Smeaton, W.A. Bethollet's *Essai de statique chimique* and its translations: a bibliographical note and a Daltonian doubt *Ambix,* 1977, **24** (3), 149–58

895 Smethurst, P.C. Gentle dissenter *Chem. and Ind.,* 1966 (4 June), 935–7

896 Smyth, A.L. John Dalton: letters to a pupil *Memoirs,* 1989–90, **129**, 139–46
Includes the texts of three letters to Mary Taylor of Moston

897 Smyth, A.L. Natural philosophy, education and industry: the Manchester contribution.
In Humby, Michael (ed.) *Libraries and information services: studies in honour of Douglas Foskett. Education Libraries Journal,* supplement 25, 1993, 129–38

898 Smyth, A.L. Some bibliographical aspects of the work of John Dalton *Manchester Review,* 1966–7, **11**, 73–9
A paper given to the Manchester Society of Book Collectors

899 Smyth, A.L. The Society's House, 1799–1940, *Memoirs,* 1986–7, **126**, 132–50
The building was always closely associated with Dalton. For over forty years he used one of its rooms as an office, study, laboratory and schoolroom. He

had the use of the main lecture room 'gratis on account of his extra-official services' and he gave 111 papers in the House to members of the Society. For a century after his death the premises housed his scientific manuscripts, notebooks, much correspondence (including his letter copy book), lecture notes and illustrations, and scientific and teaching apparatus

900 Snelders, H.A.M. The attitude in the Netherlands towards the atomic theory during the first half of the 19th century *Janus*, 1976, **63**, 111–31

901 Solov'ev, Y.I. and Petrov, L.P. Russian scientists and Dalton's atomic theory
In Cardwell, D.S.L. *John Dalton and the progress of science*, p. 300–8 **478.2**

902 Spronsen, J.W. van Dalton en Wollaston [in Dutch] *Chemisch Weekblad*, 1967, **63** (18), 217–20
Also includes a description of the Dalton bicentenary celebrations in Manchester

903 Taguchi, Roger Simple demonstrations described for Charles's Law, Boyle's Law and Dalton's Law of partial pressures *Chem. 13 News*, 1979, **102**, 14

904 Thackray, A.W. Documents relating to the origins of Dalton's chemical atomic theory *Memoirs*, 1965–6, **108**, 21–42
Using surviving original documents throws new light on how Dalton formulated and developed his chemical atomic theory

905 Thackray, A.W. Emergence of Dalton's chemical atomic theory 1801–1808 *Brit. J. Hist. Sci.*, 1966, **3**, 1–23

906 Thackray, A.W. [Entry for] Dalton, John *Dictionary of scientific biography*, 1981, **3**, 537–47

907 Thackray, A.W. Fragmentary remains of John Dalton *Ann. Sci.*, 1966, **22**, (3), 145–74

908 Thackray, A.W. 'In praise of famous men': the John Dalton bicentenary celebrations, 1966 *Roy. Soc. Notes and Records*, 1967, **22**, 40–4

909 Thackray, A.W. John Dalton: a case study in the history of science *School Sci. Rev.*, 1966, **47**, 417–25

910 Thackray, A.W. John Dalton: accidental atomist *Discovery*, 1966, **27** (Sept.), 28–33

911 Thackray, A.W. John Dalton and the beginnings of chemical atomism *Actes XIe Cong. Int. Hist. Sci.*, 1965 (publ. 1968), **4**, 63–6

912 Thackray, A.W. Medicine, manufacturers and Manchester men: the genesis of a scientific community *Actes XIIIe Cong. Int. Hist. Sci.*, 1972 (publ. 1974), **2**, 219–24

913 Thackray, A.W. Natural knowledge in cultural context: the Manchester model *American Hist. Rev.*, 1974, **79** (3), 672–709
Although not specifically concerned with John Dalton, this is an outstanding essay on the relationship between science, industry and society in Manchester and its wider implications in the Industrial Revolution. More specifically, it deals with the membership of the Lit. and Phil. of which Dalton was an officer from 1800 to 1840

914 Thackray, A.W. The origin of Dalton's chemical atomic theory: Daltonian doubts resolved *Isis*, 1966, **57**, 35–55

915 Thackray, A.W. Quantified chemistry: the Newtonian dream (In Cardwell, D.S.L. *John Dalton and the progress of science*, p. 92–108) **478.2**

916 Thackray, A.W. Science and technology in the Industrial Revolution *Hist. Sci.*, 1970, **9**, 76–89

917 Thackray, A.W. Science in one city: the Manchester story *M & B Lab. Bulletin*, 1969, **8**, 72–9

918 The Times Eye of 1844 sheds light on colour blindness *The Times*, 17 February 1995
Reports on the finding of Dalton's genetic code. He lacked the green pigment in the retina

919 Thorburn, A.D. John Dalton: his personality *The Hexagon*, 1986, **77** (2), 27–33
Up-date of an article which first appeared in *The Hexagon* in 1923

920 Trengrove, Leonard Dalton as experimenter *Brit. J. Hist. Sci.*, 1969, **4** (16), 394–8

921 Urey, H.C. Dalton's influence on chemistry (In Cardwell, D.S.L. *John Dalton and the progress of science*, p. 329–43) **478.2**

922 Voros, Laszlo Dalton (1766–1844) [in Hungarian] *Termesszettud. Kozl.*, 1966, **97** (7), 289–91

923 Waley, S.G. Dalton and chemical aspects of colour vision *M and B Lab. Bull.*, 1966, **7** (3), 42–4

924 Whitney, Bevan John Dalton and Kendal [letter] *The Friend*, 1966, **124** (37), 1100–1
History of Stramongate School 1698–1932. Claimed to be the oldest Quaker boarding school

925 Wilkinson, Lise Three drawings of Fellows by William Brockedon FRS *Notes. Rec. R. Soc.*, 1971, **26**, 183–7

Reproduces and comments on drawings of Babbage, Dalton and Faraday. See **162** and **440**

926 Wood, David A herbarium of John Dalton at the Royal Botanic Garden, Edinburgh *J. Soc. Biblphy. Nat. Hist.*, 1970, **5**, 270–1
Describes a two-volume herbarium Dalton prepared for Peter Crosthwaite. See also **104, 104.1** and **380**

927 Wren-Lewis, John The forgotten centenary *Guardian,* 8 Nov. 1966
Links Dalton with the development of plastics and the establishment of the Parkesine Company in 1866

928 Wright, W.D. The unsolved problem of 'daltonism'
In Cardwell, D.S.L. *John Dalton and the progress of science,* p. 309–28 **478.2**

?❧ Dalton in perspective – further reading

(a) Atoms and Atomists

929 Boorse, H.A., Motz, Lloyd, and Weaver, J.H. *The atomic scientists: a biographical history* New York, Wiley, 1989. 472p.
Dalton p. 34–42

930 Brock, W.H. (ed.) *The atomic debates: Brodie and the rejection of the atomic theory* Leicester, Leicester University Press, 1967. ix, 186p.

931 Knight, D.M. *Atoms and elements: a study of theories of matter in England in the nineteenth century* London, Hutchinson, 1967. vi, 167p.
Chapter 2 (16–36). Mr Dalton and his critics

932 Knight, D.M. *The transcendal part of chemistry* Dawson, 1978. viii, 289p
Nineteenth century world views on the nature of matter

933 Mellor, D.P. *The evolution of the atomic theory* Amsterdam, New York, London, Elsevier Publ. Co., 1971. viii, 171p.

934 Rocke, A.J. *Chemical atomism in the nineteenth century, Dalton to Canizzaro* Columbus, Ohio State University Press, 1984. 386p.

935 Schonland, Sir Basil *The atomists, 1805–1933* Oxford, Clarendon Press, 1968. viii, 198p.
Chapter 1. The atom in chemistry: Dalton

936 Thackray, A.W. *Atoms and powers: an essay on Newtonian matter theory and the development of chemistry* Cambridge, Harvard University Press, 1970. xxii, 326p. illus. facsims 24 cm. (Harvard monographs in the history of science)

(b) History of Chemistry

937 Bud, Robert and Roberts, G.K. *Science versus practice: chemistry in Victorian Britain* Manchester, Manchester University Press, 1984. 236p.

938 Hartley, Sir Harold *Studies in the history of chemistry* Oxford, Clarendon Press, 1971. xx, 243p.
Chapter 3 (58–91). John Dalton FRS (1766–1844) and the atomic theory (Memorial Lecture to the Royal Society, 10 November 1966) port. (R.R. Faulkner), facsim. (Dalton's notebook and symbols)
Also publ. *Proc. Roy. Soc. A*, 1967, **300**, 291–315

939 Playfair, Lyon A century of chemistry in the University of Edinburgh: being the introductory lecture to the course of chemistry in 1858.
Edinburgh, Murray and Gibb, 1858. 32p.
Dalton and Edinburgh p. 20–2

940 Russell, C.A. *Recent developments in the history of chemistry* Royal Soc. of Chemistry, 1985. x, 333p.

(c) History of Science

941 Brush, S.G. *The history of modern science: a guide to the second scientific revolution, 1800–1950.* (Iowa State University Press series in the history of technology and science) Ames, Iowa State U.P., 1988

942 Cannon, S.F. *Science in culture: the early Victorian period* Dawson, 1978. xii, 296p.

943 Cardwell, D.S.L. *The organisation of science in England.* rev. ed. Heinemann, 1972. xii, 268p.

944 Cardwell, D.S.L. *Technology, science and history: a short study of the major developments in the history of Western mechanical technology and their relationships with science and other forms of knowledge* Heinemann, 1972. xi, 244p.

945 Knight, D.M. *The age of science: the scientific world-view in the nineteenth century* Oxford, Blackwell, 1986. xii, 251p.

946 Kragh, Helge *An introduction to the historiography of science* Cambridge, Cambridge University Press, 1987. viii, 235p. bibliog.
Dalton's atomic theory p. 138–43

947 Kuhn, T.S. *The structure of scientific revolutions.* 2nd. ed. Chicago, University of Chicago Press, 1970. 210p.

948 Lipson, H.S. *The great experiments in physics* Edinburgh, Oliver and Boyd, 1968. vii, 181p.

949 Morrell, Jack and Thackray, A.W. *Gentlemen of science: early years of the British Association for the Advancement of Science* Oxford, Clarendon Press, 1981. xxiii, 592p., illus. ports. tabs. bibliog.
Includes reproduction of the Faulkner portrait, 1840. **446**

(d) Gases, Heat and Energy

950 Brush, S.G. *The kind of motion we call heat: a history of the kinetic theory of gases in the nineteenth century* Oxford, North-Holland, 1976 (repr. 1986). 2 vols. 808p.

951 Cardwell, D.S.L. *From Watt to Clausius: the rise of thermodynamics in the early industrial age* Heinemann, 1971. xvi, 336p.

952 Ewart, Peter On the measure of moving force *Memoirs*, 1813, 7, 105–258. Read 18 November 1808
Dalton appears to have been much impressed by this paper and in a letter to Johns from London, 27 December 1809, wrote 'Several have been wonderfully struck with Mr Ewart's doctrine of mechanical force. I believe it will soon become a prevalent doctrine.'

953 Fox, Robert *The caloric theory of gases: from Lavoisier to Regnault* Oxford, Clarendon Press, 1971. xv, 378p.

954 Rowlinson, J.S. The development of the kinetic theory of gases *Memoirs*, 1989–90, **129**, 29–38

See also **626 813 828–30 872 903 969**

(e) Manchester College

955 Davis, V.D. *A history of Manchester College from its foundation in Manchester to its establishment in Oxford* Allen and Unwin, 1932. 216p.

956 Manchester Academy [Proposals 1786] 'We… lamenting the dissolution of the Warrington Academy, disappointed in our expectations of its revival, and persuaded that an institute on the same liberal principles may be established in Manchester…' The Academy was founded in 1786 and the name changed to New College in 1789
MPL Reference: S&A 94/48

957 O'Brien, Padraig *Warrington Academy, 1757–86: its predecessors and successors* Wigan, Owl Books, 1989. xii, 164p., ports. illus., facsims., maps
Warrington's dissenting academy numbered among its tutors Joseph Priestley, J.R. Forster, William Enfield and John Aikin; its pupils included John Goodriche, Robert Malthus and Thomas Percival. Many members of

Cross Street Chapel, 1856, **475.4**

the Lit and Phil were alumni, including the Revd Thomas Barnes, the first principal of the Manchester Academy (later New College) which was the successor academy to Warrington. Dalton joined the College in 1793 as tutor in mathematics and natural philosophy. The classics tutor was William Stevenson whose daughter, Elizabeth, married William Gaskell, a member of the Society, in 1832

958 O'Brien, Padraig Warrington Academy, 1757–86: [alphabetical list of] students and tutors. *Memoirs*, **128**, 118–30

959 Smith, Barbara (ed.) *Truth, liberty, religion: essays celebrating two hundred years of Manchester College* Oxford, Manchester College, 1986. xxvi, 325p. illus., ports.

See also **475.2 818 884**

(f) Manchester Literary and Philosophical Society

960 Ferranti Computer Systems Manchester's men of science: the story of the Manchester Literary and Philosophical Society: being the subject of the Ferranti Computer Systems calendar, 1988. 13p. 42 x 42 cms.
High quality colour photographs with accompanying texts.

961 Heidmann, W. *Die pädagogischen Bemühungen im Umkreis der Manchester Literary and Philosophical Society im späten 18. Jahrhundert* Brunswick, Technical University, 1987. 119p.

962 Makepeace, C.E. *Science and technology in Manchester: two hundred years of the Lit. and Phil.* Manchester, Manchester Literary and Philosophical Publications Ltd., 1984. 64p., ports., illus., facsims.
An illustrated history of the Lit. and Phil. (115 illustrations) showing the scientific and technological achievements of members of the Society

963 Manchester Literary and Philosophical Society *Complete list of the members of the Society from its institution on February 28th, 1781 to April 28th, 1896, and bibliographical lists of the manuscript volumes dealing with the affairs of the Society, and of the volumes of the* Memoirs and Proceedings *published by the Society, with two appendices.* Manchester, The Society, 1896. 53p.
Appendix 1. Rules established for the government of the Society and a list of members. 1782.
Appendix 2. A short account of the institution and views of the Society. [1783?]

964 Manchester Literary and Philosophical Society *Index to the* Memoirs and Proceedings, *1781–1989* Manchester, Manchester Literary and Philosophical Publications Ltd., 1991. xvi, 373p.
An author-subject index to the 154 volumes published between 1785 and 1990, containing approximately 12,000 entries (of which 100 are under 'Dalton'). Dalton was responsible, as Secretary, for editing the *Memoirs* from 1800 to 1808 and because he felt that literary papers were of little value 'as contributing no positive facts to our stock of knowledge, and in short producing nothing', the volumes for many years became almost exclusively concerned with science and its applications. It should be noted that some authors and their contributions had an important influence on him

See also entries under 'Manchester Literary and Philosophical Society' in the Subject Index and appropriate entries in the Index to the *Memoirs*.

(g) Manchester – Science, Technology and Industry

In a populous town like this, where the arts and manufactures are so intimately connected with various branches of science, it may be presumed that public encouragement will not be wanting… John Dalton – Prospectus of lectures on natural philosophy, Manchester 1805

965 Brooks, Ann, and Haworth, Bryan *Boomtown Manchester, 1800–1850: the Portico connection* Manchester, Portico Library, 1993. xii, 136p.
'A history of the Portico Library and Newsroom and the influence of the

Secretary's room in the Manchester Literary and Philosophical
Society's house at 36 George Street, which was used by Dalton as a laboratory,
*c.*1915 **474·5**

Exterior of the Manchester Literary and Philosophical Society's house at 36 George Street, 1904. Dalton used the room to the left of the entrance 474·7

founding members on the development of Manchester'. Many members of the Society were also members of the Portico Library. John Dalton was the first honorary member

966 Burkhardt, G.N. The school of chemistry in the University of Manchester *J. Inst. Chem.*, 1954, 448–60

967 Butler, Stella *Atoms, energy and industry: two centuries of Manchester science* Manchester, Manchester Museum of Science and Industry Trust, 1985. iv, 27p.
Published for the 30th Congress of the International Union of Pure and Applied Chemists

968 Cardwell, D.S.L. (ed.) *Artisan to graduate: essays to commemorate the foundation in 1824 of the Manchester Mechanics' Institution, now in 1974 the University of Manchester Institute of Science and Technology* Manchester, Manchester University Press, 1974. viii, 284p.

969 Cardwell, D.S.L. *James Joule: a biography* Manchester, Manchester University Press, 1989. x, 333p.
Joule, Dalton's most famous pupil, was a 'provincial' scientist like his mentor. It has been said of him that the Society was his inspiration, his university and his club. He achieved the distinction of giving his name to the SI unit of work and energy

970 Cardwell, D.S.L. Manchester in the 19th century: a case study in the patronage of science. Typescript, 1972. 22p.
Paper presented to a joint meeting of the Past and Present Society and the British Society for the History of Science held at the Royal Society in London on 14 April 1972
MPL

971 Chaloner, W.H. Manchester in the latter half of the eighteenth century *John Rylands Library Bulletin*, 1959–60, **42**, 40–60

972 Field, Clive, and Pickstone, J.V. (eds.) *A centre of intelligence: the development of science, technology, and medicine in Manchester and its University.* Manchester, John Rylands University Library, 1992. 48p.

973 Hills, R.L. A monument to the Industrial Revolution: the North Western Museum of Science and Industry *Memoirs*, 1993–4, **132**, 123–47
The successor museum is the Museum of Science and Industry in Manchester

974 Kargon, R.H. *Science in Victorian Manchester: enterprise and expertise.* Manchester, Manchester University Press, 1977. xi, 283p.
The most extensive contribution to the subject. Many references to Dalton and the Society

975 Lipson, H.S. The *fundamental* fundamental particles *Memoirs*, 1987–8, **127**, 115–6
'There are so many so-called fundamental particles but three are most important – the proton, the neutron and the electron. Is it generally realised that all three were Mancunian in origin?'

976 Lipson, H.S. The last fifty years of physics in Manchester *Memoirs*, 1980–81, **121**, 87–99

977 Musson, A.E. and Robinson, Eric *Science and technology in the Industrial Revolution* Manchester, Manchester University Press, 1969. viii, 534p.
Much of the book is devoted to the Manchester region

978 Pickstone, J.V. Some Manchester sources for the history of science, technology and medicine, with special reference to the John Rylands University Library *John Rylands Library Bulletin*, 1989, **71** (2), 141–57
An essay on the various Manchester institutions and their holdings in the fields indicated by the title. Useful for its evaluation of individual collections especially in bringing to notice rare and unique material

979 Rutherford, Ernest The scattering of the alpha and beta rays and the structure of the atom *Proceedings*, 1911, **55**, 18–20
Rutherford first announced the 'splitting' of the atom to a meeting of the Society on 7 March 1911, from the same platform used by Dalton to propound his atomic theory on 21 October 1803

980 Wetton, Jenny Scientific instrument making in Manchester, 1790–1870 *Memoirs*, 1990–1, **130**, 37–68

981 Williamson, Raj Arthur Schuster: pioneer in physics teaching *Memoirs*, 1988–9, **128**, 30–45

982 Williamson, Raj H.G.J. Moseley: the Manchester School and the atom *Memoirs*, 1987–8, **127**, 66–79

983 Williamson, W.C. *Reminiscences of a Yorkshire naturalist* 1896 repr. J. Watson and B.A. Thomson, Williamson Bldg., University of Manchester, 1985. xxii, 228p.

Chapters 4 and 5 deal with Dalton and the Society

984 Wyke, T.J. Nineteenth century Manchester: a preliminary bibliography In Kidd, A.J. and Roberts, K.W. *City class and culture* Manchester University Press, 1985)
Lists 1200 items arranged in 27 subject groups, e.g. 5 – textiles, 6 – engineering and metal industries, 8 – other industries, 11 – railways, 16 – public health and hospitals, 19 – education, 20 – science

See also **528 532 565 566 810 812 912 913 917** and appropriate entries in the Index to the *Memoirs* **964**

(h) Quakerism

985 Blamires, David *Quakerism and its Manchester connexions:* an exhibition held in the John Rylands Library, 6 February–23 May 1991. Manchester, MU, 1991. 28p.

986 Curwen, J.F. *Kirkbie-Kendall* Kendal, T. Wilson, 1900. 455p.
Provides useful information on the town of Kendal in Dalton's time there. The Friends Day School, Stramongate (founded 1698) is described. p. 402–5

987 Foulds, E.V. and others *Mount Street, 1830–1930: an account of the Society of Friends in Manchester, together with short essays on Quaker life and thought,* written for the Centenary of the Friends Meeting House, Mount Street, Manchester. Manchester, Mount Street Centenary Committee, 1930. 53p.

The Friend's Meeting House, Mount Street, Manchester,
*c.*1850 **475.6**

988 Foulds, E.V. *The story of Quakerism* Bannisdale Press, 2nd ed., 1960. 310p.

989 Irwin, Margaret *The history of Pardshaw Meeting and Meeting House* Society of Friends, [1919] 28p.
Includes illus. of Friends Meeting House, Eaglesfield, used by Dalton as a school

990 Jones, R.M. *Later periods of Quakerism* MacMillan, 1921. 1020p.
Dalton p. 762–8

991 Raistrick, Arthur *Quakers in science and industry: being an account of the Quaker contributions to science and industry during the seventeenth and eighteenth centuries* Newton Abbot, David and Charles, 1968. 361p. (first publ. Bannisdale Press, 1950)

992 Stewart, W.A. Campbell *Quakers and education as seen in their schools in England* Epworth Press, 1953. 320p.
p. 6–81 deal with education to 1840

See also **144 167.22 356–356.6 366.1 457.2 475.6 499 505 527 570 577 808 868 869 883 924**

(i) Colour Vision

993 Dickinson, C., Murray, I.J. and Carden, D. (eds.) *John Dalton's colour vision legacy:* proceedings of an international colour vision conference held in Manchester, September 1994. Taylor and Francis, 1996 [in the press] See also **392.7**

994 *Dalton and the contribution of self observation to scientific discovery: an account of colour blindness* Roseneath Scientific, c. 1992 B93–14341 18p.

995 Sherman, P.D. *Colour vision in the nineteenth century* Bristol, Hilger, 1981. 260p.

996 Wilson, George *Researches in colour blindness: with a supplement on the danger attending the present system of railway and marine coloured signals* Edinburgh, Sutherland and Knox, 1855. xx, 175p.
p. 160–1 On the terms 'Daltonism' and 'Daltonian'

997 Rosetta Pictures for BBC tv The mind traveller – *1. The island of the colour blind* broadcast 14 November 1996, running time 45 minutes, 1996
Presenters – Oliver Sacks, Knut Norby; Executive Producer – Robin Brightwell; Producer – Emma-Crichton-Miller; Director – Christopher Rawlence
Investigates colour blindness in the Pacific island of Pingelap where one in twelve of the population is affected. Considers early theories of colour blindness and shows Dalton's eye. Proposes that colour perception is not with the eye but the brain
See also **392.7 472.4 820 822 844 853 875 918 923 928 1010**

998–9 not used

Part Three ❧ DALTON'S SURVIVING APPARATUS AND PERSONAL EFFECTS

❧ Scientific objects

Dalton used a room in the Society's House as a laboratory from 1800 until his death in 1844. He bequeathed his physical and chemical apparatus to W.C. Henry who, on Dalton's decease, left the collection with the Society. It did not therefore leave the building but was stored in boxes in a back room and was somewhat neglected until, in 1857, money became available for the apparatus to be 'examined and cleaned by competent persons;' and placed 'in a handsome oaken case with glass doors'.[1] The first published listing of any part of the collection was in 1876 when it was exhibited at South Kensington and some 52 items were enumerated and described.[2] Electrical apparatus, models of mechanical powers, models of steam engines, air pumps and other apparatus of a similar kind were purposely excluded.

In anticipation of the centenary of Dalton's atomic theory in 1903, the

The show-case of Dalton's apparatus, c.1915 **474.9**

most interesting pieces were placed in cases in the Library and Council resolved that photographs of the apparatus were to be taken for reproduction in the *Memoirs*. This resulted in nine pages of plates and a descriptive article by Francis Jones appearing in volume 48.[3] In connection with the British Association meeting in Manchester in 1915, a pamphlet was published describing the Society's House and its contents which, of course, included many Dalton items then specially displayed.[4] The opening of the Thomas Thorp Room in February 1936 provided the opportunity of exhibiting 'the large collection of scientific apparatus and other relics accumulated' since the Society's institution. A descriptive leaflet included a further listing of Dalton items.[5] Much of the collection was destroyed in an air raid on 24 December 1940; a list of the articles salvaged was published in 1942.[6]

A small collection of Dalton items was owned by the Society of Friends and kept in Dalton Hall (established in 1876), a students' hall of residence.[7] Most of this had been given by the family of Isabella Benson who with her sister Hannah were probably Dalton's nearest relatives at the time of his death. A few objects were given to the Science Museum (London) in 1949 and a small number retained in the Hall but the major part of the collection was presented to the Society by the Friends in 1958.[8]

With the opening of the new building in 1960, the surviving apparatus was once more displayed in the Society's premises, this time in wall cases in the Dalton Room on the first floor, whilst the Clegg beam engine was placed in a prominent position in the entrance hall. The building was the main venue for many of the activities connected with the Dalton Bicentenary celebrations in 1966. A special exhibition was held in the Manchester Museum (Manchester University), with an accompanying descriptive pamphlet,[9] and an international conference of historians of science took place. The published papers of this conference include a well illustrated contribution by Kathleen R. Farrar which is by far the most extensive descriptive account of Dalton's apparatus and is also a firm rejection of Roscoe's 'ink-bottle legend'.[10] Dr Farrar emphasizes that much of the apparatus was obtained for teaching rather than research and, as teaching was his livelihood, Dalton was prepared to spend liberally on equipment as a business investment.

References

1. Annual report of Council, 1857–58. *Proceedings*, 1857–60, **1**, 53–4

2. **Roscoe, H.E.** Notes on a collection of apparatus employed by Dr Dalton… which is about to be exhibited… at South Kensington. *Proceedings*, 1875–6, **15**, 77–82

3. **Jones, Francis.** The collection of apparatus used by Dalton, now in the possession of the Society. *Memoirs*, 1904, **48**, (22), 1–5

4. **Barnes, C.L.** The Society's House, *Proceedings*, 1915–6, **60**, 2–8

5. Manchester Literary and Philosophical Society. *The Thomas Thorp Room.* 1936. 2p.
LP

6. List of articles salvaged from 36 George Street, Manchester, after the destruction
of the building on 24 December 1940 *Proceedings*, 1939–41, **84**, 34–6

7. **Marsh, J.T.** Old Quaker Dalton. *Memoirs*, 1968–9, **111**, 42–3

8. Dalton relics formerly belonging to Dalton Hall. Typescript, 3p. [1958]
LP

9. Manchester Museum. *The Dalton exhibition, 1966.* Manchester, Manchester Museum,
1966. 7p.
LP

10. **Farrar, K.R.** Dalton's scientific apparatus (in Cardwell, D.S.L. *John Dalton and the
progress of science*, 1968, p. 159–86)

The Museum of Science and Industry in Manchester, Liverpool Road, Castlefield, Manchester M3 4JP

(Except **1013**) For further information and access to the collections, ring
0161 832 2244; fax 0161 833 2184
* – formerly in Dalton Hall

1000 *Thermometer* Broken with bulb missing, wooden backplate with two
scales – 10 to 140 degrees Fahrenheit and 0 to 50 degrees Réaumur. Two
holes for mounting screws at top, one at bottom. Used by Dalton at his
house in Faulkner Street
Size: 4 x 2.5 x 27cms.
With this you will receive an old thermometer which formerly belonged to
the late Dr Dalton and is the one he placed outside his window in Faulkner
St for many years; it was purchased at the sale of the late Peter Clare's effects
– Joseph Manchester to John Harland, 17 March 1853 **368.5**

1001 *Barometer* Mercury in glass tube barometer on red wood backplate with
wood cover to reservoir bulb. Printed paper scale with likely weather
conditions on one side, graduated 28 to 31 on the other. Made by Joshua
Ronchetti, Manchester
Size: 10 x 6.5 x 90 cms.

1002 *Water thermometer bulb* Bulb made of china clay, blackened on one side,
base flattened
Size: 4.5 x 5 cms.

1003 *Glass phial* Tall necked glass phial, cracked at bottom. Used for mercury
amalgams
Size: 2.5 x 7.5 cms.

1004 *Glass bottle* Parallel sides, badly cracked, ground glass hole for stopper at top, bottom missing
Size: 3 x 9 cms.

1005 *Eudiometer* Glass spark eudiometer with no metal fittings, a weighted bulb with holes at either end
Size: 5 x 14 cms.

1006 *Mugs* Three earthenware mugs with handles, originally brown but blackened in fire. Possibly used for holding mercury
Size: 7 x 9 x 7.5 cms.

1007 *Thermometer backplate* Wooden plate with two hanging holes and tube holder at each end (tube missing)
Size: 26.5 x 2 x 9 cms.

1008 *Model of a Watt Beam Engine* A double-acting single cylinder beam rotative engine. The beam operates a flywheel through a connecting rod, crankshaft and a 2:1 increasing gear. The condenser and air pump are situated in a copper tank below and adjacent to the cylinder, the air pump plunger being operated by the beam. The inlet and exhaust valves are of the poppet type with co-axial stems and the rocker arms provide positive opening and closing. The valve mechanism is operated by an eccentric, keyed to the main shaft by wooden wedges, which in turn operates a rocking lever carrying the push rods to the rocker arms. On the brass cylinder is engraved 'Clegg, David Street, Manchester'. The engine appears to be a working model and a running test on steam was achieved after the thorough renovation in 1962–3. A schedule of repairs and replacement of missing parts accompanies the model
Size: 76 x 35.5 x 71 cms.
Dalton is known to have used this model in his lectures. Samuel Clegg, the gas engineer, was a pupil of Dalton's after which he was apprenticed with Boulton and Watt where his name is recorded as being in 1802. He returned to Manchester in 1806 and after this time it seems that he probably presented his old teacher with the model. The latter is mentioned a number of times subsequently and it appears on the list of articles salvaged after the destruction of the Society's premises. The description of the engine is based on an article by the two men mainly responsible for its renovation – A.P. Hatton and J.W. Flowett. Clegg's model of a Watt beam engine. *Manchester Memoirs*, 1963–4, **106**, 104–7

1009 *Notebook* Hard-backed exercised notebook – 'Dalton Apparatus; by W.W.H. Gee
Size: 21 x 16.5 cms.

1010 *Eye* Two parts of Dalton's eye: smaller part (probably) lens or detached retina; larger part (probably) eye cup. Kept in contemporary watchglasses into which they were put after dissection
Size: 5 cms.
Dalton's explanation of his colour blindness was 'that one of the humours of my eye… is a coloured medium, probably some modification of blue. …I apprehend it might be discovered by inspection'. J.A. Ransome, his close friend and medical adviser, made a careful post mortem examination of the eye; 'the *aqueous* was collected in a watchglass… [and] was found to be perfectly pellucid and free from colour'. Using a minute part of the eye, Dalton's genetic code was determined in 1994. The DNA analysis showed him to have been a deuteronome. See **853**   *

1011 *Atomic Models* Five wooden balls used by Dalton for demonstrating atomic theory. Three are joined together, two separate
Sizes: 3 and 2.5 cms.   *

1012 *Atomic Models* Three wooden balls used by Dalton to expound his atomic theory. Small holes are bored in each so that they can be joined together to form ball and spoke models to illustrate the building blocks of molecules. The balls were made for Dalton by his close friend and fellow member of the Society, Peter Ewart, an engineer and forerunner of Joule in the study of energy.
'My friend, Mr Ewart, at my suggestion, made me a number of equal balls, about an inch in diameter, about thirty years ago; they have been in use ever since, I occasionally show them to my pupils. One ball had 12 holes in it, equidistant; and 12 pins were stuck in the other 12 balls so as to arrange the 12 around the one and be in contact with it; they (the 12) were about one-tenth of an inch asunder. Another ball, with 8 equidistant holes in it; and they (the 8) were about $3/_{10}$ of an inch asunder, a regular series of equidistant atoms. …I had noticed at that time (30 years ago) that the atoms were not all of a *bulk* but for the sake of illustration I had them made alike.' – *On a new and easy method of analysing sugar*, 1840, p. 3–4
SM   *

1013 *Microscope slides* (12) Made by Dalton
BONE:
7 slides 6.5 x 1.2 cms. each with 4 specimens
2 slides 7.0 x 1.2 cms. each with 4 specimens
2 slides 13.0 x 1.7 cms. each with 6 specimens
BRASS: 1 slide 9.0 x 1.3 cms. with 4 specimens
Manuscript list of the specimens 32.5 x 11 cms.
LP   *

Manchester University Chemistry Department Collection

Examined by Kathleen Farrar (ref 10 p. 135 above); photographs of the collection are on two plates, 2b and 3b, in her article.

Plate 2b
1014 *Cubicle flasks* (3) Purpose not known but similar ones in Oxford Museum of the History of Science (p. 173–4)
MU

1015 *Water thermometer* Used to determine the temperature at which water exhibited its maximum density (p. 173)
MU

1016 *Bottles* (2) About 8oz capacity, lined inside and out with metal foil. Probably the remains of improvised Leyden jars (p. 174)
MU

1017 *Specific gravity bottle* Dalton appears to have relied on specific gravity for standardizing his solutions (p. 180–1)
MU

1018 *Thermometer* (p. 171–2)
MU

Plate 3b
1019 *Mountain barometer* Tube broken and part of the case missing (p. 163–70) A very similar barometer signed by Laurence Buchan, Manchester, c. 1835, is in the National Museums of Scotland. Dalton's barometer may have been made by Buchan, who was a fellow member of the Society
MU

1020 *Mountain barometer case* Empty case with ferule missing (p. 163–70)
MU

Manchester Grammar School

Examined by Kathleen Farrar (ref 9 above). Francis Jones, author of the 1904 article (ref 3 above), was a former chemistry master at the school

1021 *Round Flask* Brass fitted, rather less than a litre capacity (p. 183)
MGS

1022 *Barometer tubes* (2) Similar to that in Dalton's barometer made by Ronchetti (p. 183)
MGS

1023 *Cradle* Dark wood (probably oak) child's cradle on rockers (one at foot renewed). At head end the 'roof' cover has removable top, panelled ends, knob at one end corner missing
Size: 94 x 47 x 67 cms.   *

1024 *Umbrella* Opening umbrella with eight ribs (possibly whalebone) mounted on cane centre pole, with black iron slider and supporting ribs. Made by D. Radford, Market Street, Manchester
Size unopened: 5 x 91.5 cms.   *

1025 *Lock of Dalton's hair* Wrapped in paper on which is written 'Dr Dalton's Hair'
'Hair cut from the side of the front of head of John Dalton DCL FRS etc. etc. on Tuesday, May 2nd 1837, and presented to me by his housekeeper Miss Wood on Monday, May 7 1837, at Dr Dalton's house. Henry Hough Watson NB Dr Dalton was attacked with paralysis of his right side on Tuesday morning April 18th 1837 about 8 o'clock'
MU   *

1026 *Lock of Dalton's hair* Wrapped in paper on which is written: (outside) 'Dr Dalton's hair, cut off Jany 2nd 1840', (inside) 'Elizabeth Frankland'   *

1027 *Spectacles* Pair of tortoiseshell folding arm spectacles with one lens missing
Size: 11.5 x 2.5 x 17 cms.   *

1028 *Calling card and plate* Engraved copper printing plate and calling card
Size of card: 7.5 x 4 cms.   *

1029 *Ivory umbrella handle* Engraved 'Dr Dalton'
Size: 7 x 2 cms.   *

1030 *Cap* Lined cloth cap worn by Dalton when working (his thinking cap)
Size: 34 x 13.5 cms.   *

1031 *Razor* Cut-throat razor in leather covered papier mâché case, engraved 'Richardson'
Size: 25 x 2 x 1 cms.   *

1032 *Gold watch and key* Inscribed 'To J.D. from P.C.' Given to Dalton by Peter Clare and presented to the Society by Miss Agnes Wood in 1934. The watch survived the 1940 fire in the office safe.
Peter Clare (1781–1851) was a clock, watch and barometer maker. The most important example of his skill is the 'New Time-Keeper and Recorder' made for the Town Hall, then in King Street. It is 7 feet high and is notable for a

mercury compensatory pendulum. It is now displayed in the City Art
Gallery. He was asked by Dalton to make a clock for the Society which
would strike only once in twenty-four hours, at 9pm, with the object of
closing meetings at that time. The metal parts of this clock were recovered
after the fire and left with a clock repairer but never retrieved and it is now
lost.

Clare was elected to the Society in 1810 and became a vice-president. A
Quaker, he was one of Dalton's closest friends and the main executor of his
will

LP

1032.5 *Chair* With Dalton Collection
MU

1033 *Gravestone* Polished red granite slab with the words JOHN DALTON.
Size: 183 x 51 cms.

In 1853, a testimonial public subscription raised a sum of £5,312, part of
which was spent on a granite monument over Dalton's grave in Ardwick
Cemetery. This was a plain but substantial construction enclosed by iron
railings. A massive block was surmounted by an overhanging slab, slightly
pitched, and this surviving flat slab on which was the name JOHN
DALTON. The tomb was restored by the Society in 1899 (*Proceedings*, 1899,
44, 2, 29, 50) and renovated again in 1945 (*Proceedings*, 1945–6, **87**, 8).
Overgrown and neglected, the cemetery was closed in 1950; delayed by legal
problems, the gravestones were eventually cleared and the land used for a
children's recreation ground, opened in 1966. The slab with Dalton's name
was built into the wall of the entrance hall of the Society's new building and
the base of the monument set by the side of the Theed statue **472** in the
fore-court of the then new John Dalton College of Technology (now the
Faculty of Science and Engineering Building, Manchester Metropolitan
University), Chester Street. When the Society's premises were demolished in
1980, the slab was transferred to the Friends Meeting House in Mount Street
(where it sat somewhat uneasily against a wall) and from there to the Science
Museum in Castlefield.

The site of Ardwick Cemetery is off Hyde Road, between Dalberg St and
Devonshire St North; the old entrance was at the end of Ford St.
See also **476.7** and **869**

Medals

1034 Gold
Obverse GEORGIUS IIII REX SOC. REG. LOND. PATRONUS 1826
Reverse REGIS MUNIFICENTIA ARBITRIO SOCIENTATIS. NEWTON
Rim JOHN DALTON MDCCCXXVI

In 1825, George IV announced that he was establishing the award of two gold medals, worth 50 guineas each, to be given annually by the Royal Society. Accordingly, in 1826 the Council of that Society decided that the medals should be awarded for the most important contributions to science and that **John Dalton** would be the first recipient for his discovery of the Atomic Theory. However, the money for the medals was not forthcoming until the reign of William IV and then the medals were presented retroactively. (See **Humphry Davy**. Presidential address on the occasion of the presentation of the first Royal Medal of the Royal Society to John Dalton. *Collected works*, 1839, 7, 92–9)
LP

1035 Silver
Obverse JOHN DALTON DCL FRS
Reverse STRUCK IN COMMEMORATION OF THE MEETING OF THE BRITISH ASSOCIATION HELD IN MANCHESTER AND IN HONOUR OF DR JOHN DALTON BY THE PROPRIETORS OF BRADSHAW'S JOURNAL JUNE 1842
LP

1036 Replica of the above
LP

1037 Bronze
Obverse PRESENTED BY THE LITERARY AND PHILOSOPHICAL SOCIETY OF MANCHESTER (Head of Dalton)
Reverse THE SOCIETY INSTITUTED FEBRUARY 28TH, 1781
The Dalton Medal
A silver medal is presented
Recipients
1898 Edward Schunck, FRS
1900 Henry E. Roscoe, FRS
1903 Osborne Reynolds, FRS
1919 Ernest Rutherford, OM, FRS
1931 Joseph J. Thomson, OM, FRS
1942 Lawrence Bragg, CH, MC, FRS
1948 P.M.S. Blackett, OM, FRS
1966 Cyril Hinshelwood, OM, FRS
1981 Dorothy M.C. Hodgkin, OM, FRS
(Six of those named had been Ordinary Members)
LP

Part Four ❧ DALTON'S MANCHESTER

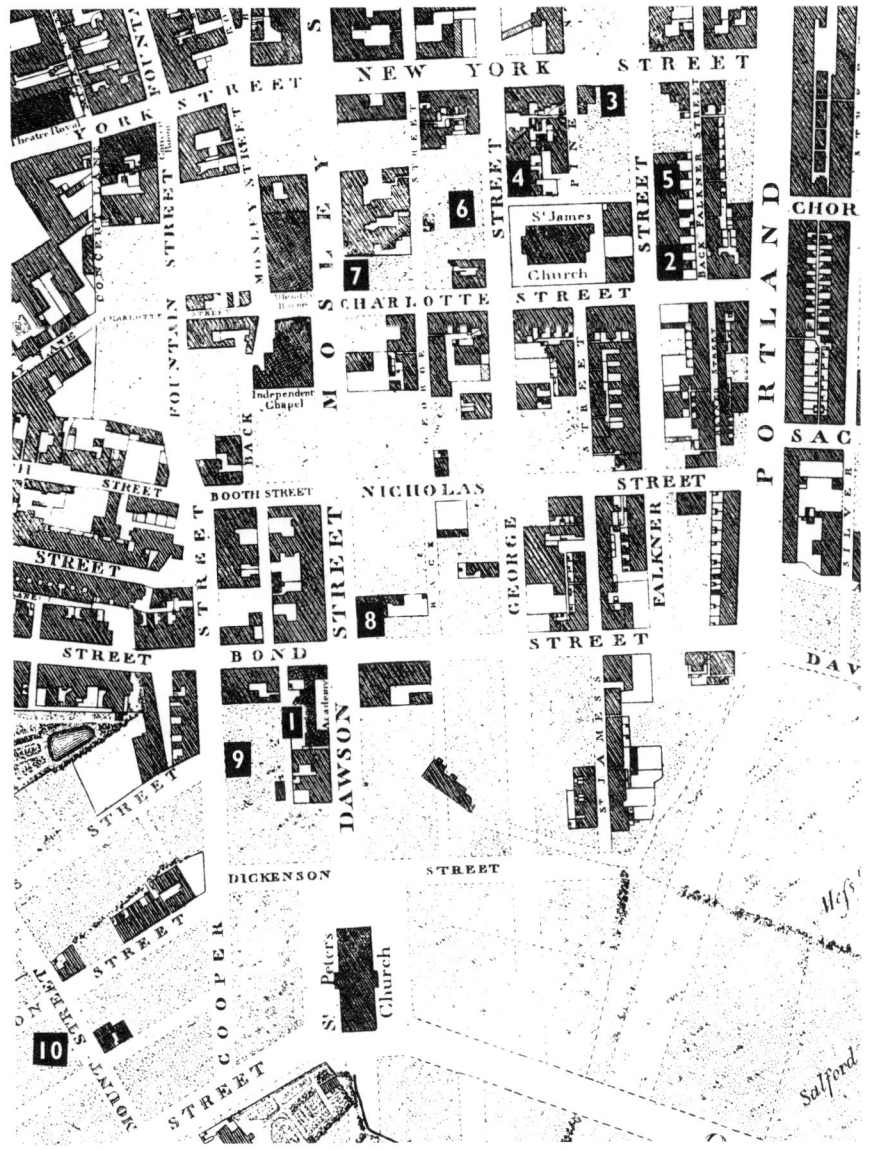

Manchester in 1793 *A town in transition*

1038 Green, William

A plan of Manchester and Salford drawn from an actual survey by William Green, begun in the year 1787 and completed in 1794. Engraved by J. Thornton. Manchester, 1794. 116 x 143 cms. Scale: 24 inches to 1 mile (1:2,640)

Green's plan is the first large scale survey of Manchester and Salford and aims to 'lay down all the streets, squares, spaces, courts, passages, fields, gardens, etc, in such a manner as to express the exact dimensions of every regularly bounded plot of land'. It also shows the names of the respective owners of open land and proposed new streets and canals are delineated. This survey is of special interest as its date not only coincides with Dalton's arrival in Manchester but also the beginnings of the Industrial Revolution when land for industry, communications, commerce and housing became increasingly in demand. The medieval town, centred mainly on Market Street Lane, Dean's Gate, Smithy Door and Old Mill Gate, was about to be largely demolished, whilst the new industrial premises – dye works, mills, factories, foundries, printing works, breweries, cotton works – already feature on the plan. 'Mr Drinkwater's cotton works', the first Lancashire cotton mill to be powered by Boulton and Watt engines in 1789, is shown to be in Auburn Street, near Piccadilly. Less than half a mile away in Dawson Street (now part of Mosley Street) is another innovation, the Manchester Academy (opened in 1786, name changed to New College, 1789).

In 1793, Dalton came from Kendal to Manchester to take up the post of tutor at New College which stood on land in the south of the town, sold by the Lord of the Manor in 1783. The neighbourhood had become a fairly affluent residential district through which ran the aptly named Mosley Street from the Infirmary to Dawson Street and St Peter's Church. Some notable buildings were to be erected in the area – the Assembly Rooms, Portico Library, Royal Manchester Institution (Mosley Street); the Athenaeum (Bond Street); the Friends Meeting House (Mount Street); and, most important of all for Dalton, the Literary and Philosophical Society (George Street). Dalton's name was proposed for membership by Robert Owen, Drinkwater's able young manager, in 1794. Peter Drinkwater was also a member; both men were shrewd industrial innovators, typical of many of the members.

The opening of the Rochdale canal in 1804 attracted numerous workshops and factories which began to hem in the Mosley Street area and, as the less pleasant aspects of living in an industrial town became more apparent, most residents moved further out to the countryside. Many of the old Georgian terraces were replaced by textile warehouses and offices but Dalton stayed on in his house in Faulkner Street until his death in 1844.

See also **472.5**

Dalton's Manchester

Numbers 1–10 appear on the 1794 plan of Manchester, numbers 1–18 on the 1997 plan of Manchester.
* – now demolished. Numbers in bold refer to entries in the main text

*1. New College, Dawson St (now Mosley St). Site near corner of Princess St is now a garden. Tutor 1793–1800, lodgings 1793–8, **818, 884, 955–9**

*2. 35 Faulkner St, rooms, 1798

*3. 18 Faulkner St, lodgings with John and Mary Cockbain, 1798–1804

*4. 25 George St. Lodged with the Revd and Mrs William Johns, 1804–30. When the Johns family moved, stayed on until 1834 with a housekeeper

*5. 27 Faulkner St (opposite Chain St). His own house, 1834–44, **475**

*6. Lit and Phil, 36 George St 1799–1940. Site now Devonshire House (plaque). Here, Dalton had his office, laboratory and schoolroom; his scientific papers and apparatus, as well as other mss., were stored in the Society's house until its destruction in an air raid in December, 1940, **476.5, 899, 960–4**

7. Portico Library, Mosley St 1806– . Subscription library where Dalton came to read the newspapers, usually twice a day. He was its first honorary member, **475.2, 965**

8. Royal Manchester Institution, Mosley St 1830– , now City Art Gallery. Dalton gave course of lectures here on meteorology in 1834. Clare's 'time-keeper and recorder' on display on main staircase, **378, 465, 1032**

*9. Mechanics' Institution, Cooper St (opposite Lloyd St) 1825–57. First building erected for the purpose in England. Dalton gave a course of five lectures here on meteorology and one on the atomic theory in 1835. He was a 'loyal and sympathetic supporter' of the Institution and was its first Vice-President. UMIST is directly descended from the Institution, **362.4, 378, 968**

10. Friends Meeting House, Mount St 1830– . Dalton was a member of the Manchester Meeting from January 1794 and a subscriber to this building. It was used for the 1842 British Association Meeting, **474, 475.6, 869, 987**

11. Manchester Town Hall 1877– , Albert Square. Dalton's statue (Chantrey), 1837 stands at the main entrance together with that of his pupil, Joule. First floor – Madox Brown mural (Main Hall), Faulkner portrait (oils) (Reception Room), **441, 447, 465**

*12. Cross St Chapel 1694–1940. The Lit and Phil had its meetings in a room (which survived until 1995) at the back of the Chapel, 1781–99. Dalton gave his first six lectures to the Society here, **475.4**

13. Chetham's Library, Long Millgate. Founded 1653 (building dates from 1421). The oldest public library in the English speaking world. Dalton regarded this library as most important to his studies, **475.9**

14. Rylands Library, Deansgate 1899– . A world renowned collection which includes most of Dalton's surviving papers, **481.7**

15. Museum of Science and Industry, Liverpool Rd, Castlefield. The exhibits illustrate much of the area's technological and industrial history, including

the first passenger railway station. Most of Dalton's surviving apparatus and personal effects are here, **1000–1033**

16. Manchester Metropolitan University, forecourt off Chester St. Theed statue 1854, formerly in Piccadilly, **472, 1033**

*17. Dog and Partridge Inn, Chester Rd, Old Trafford. Original building a 16th-century farm; became an inn late 18th century. Rebuilt 1900. Dalton regularly played bowls here, **475.8**

*18. Ardwick Cemetery, Devonshire St North, Ardwick. Dalton buried here in 1844. The gravestones were cleared on the closure of the Cemetery in 1950 and the site subsequently used as a children's recreation ground, **467.7, 1033**

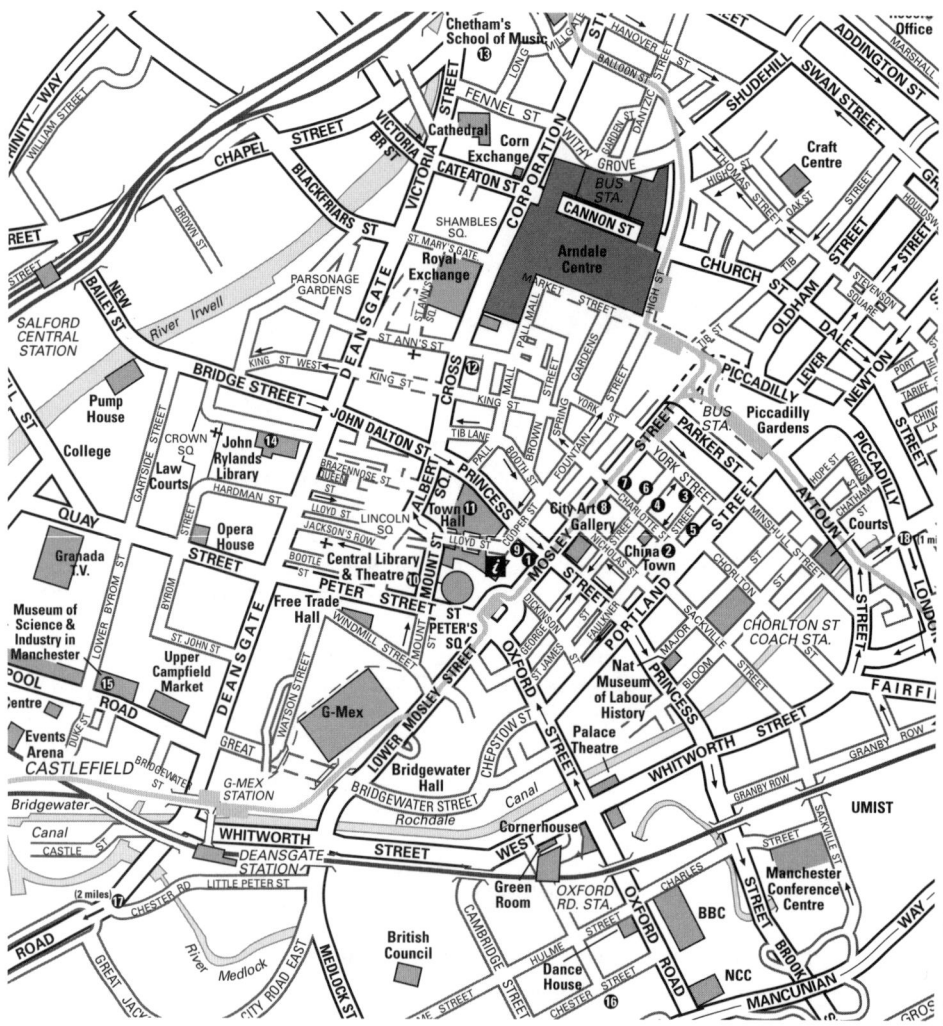

?⬦ Index of letters and printed texts of letters

No detailed record exists of the Society's large collection of Dalton correspondence held before 1940 when the major part of it was destroyed. Other letters have become widely dispersed over the years and the locations of many are not generally known. Fortunately, some of the texts of lost letters have survived in the various biographies of Dalton, although often quoted only selectively. It might be necessary, therefore, to refer to more than one printed source to build up the text of a particular letter. Furthermore, Dalton sometimes made a copy letter which is not identical with that received by the correspondent (e.g. Mary Taylor, 13 June 1825, Brockbank quotes the copy letter)

Letters are listed alphabetically by name of correspondent in three separate sequences:

1. from Dalton. 2. to Dalton. 3. between third parties relating to Dalton.

Biographies containing letters are indicated by the following abbreviations, followed by the page numbers of the letter or part of the letter concerned. The bold number is the running number in the Bibliography.

B – Brockbank (1944)	P – Patterson (1970)
C – Cardwell (1968)	R – Roscoe (1895)
G – Greenaway (1966)	RH – Roscoe and Harden (1896)
H – Henry (1854)	S – Smith (1856)
L – Lonsdale (1874)	T – Thackray (1972)
M – *Manchester Memoirs* (vol. no., then page)	

1. Letters from Dalton

Abbatt, H.	13 Apr. 1839	**155**	B28
(see also Tipping, H.)	14 Sep. 1842	**156**	B28
Alderson, W.	4 Aug. 1788		L57–60, R34–5
Arago, F.J.D.	4 May 1836		H21–2
	13 Nov. 1836	**156.2**	
Babbage, C.	15 May 1830	**157**	T160
	7 Dec. 1830	**158**	T160–1
Backhouse, J.	20 Sep. 1799	**159**	T noted
	23 Mar. 1801	**159**	T noted
Bernard, T.	14 Jul. 1804	**159.2**	M108, 33–4; T106–7

(Opposite) Places with Dalton associations in present-day Manchester
© Marketing and Visitor Services, Manchester City Council

* Twenty-five letters to Henry Dalton, 1834–42, mainly concerned with family property, are in the County Record Office, Carlisle – see Thackray **492.5** p. 128

2. Letters to Dalton

Spring-Rice, T.	24 Nov. [1834] **349**			
Stanley, E.J.	6 Jun. 1836		R201	
Suliot, T.E.	3 Mar. 1837	**349.1**		
	16 Mar. 1837	**349.1**		
Taylor, C.	17 Sep. 1838	**350**		
Thomson, C.P.	22 Jun 1833		H180–1; R201–2	
	8 Jun. 1836		H181	
Thomson, T.	8 Mar. 1807		RH141–3	
	13 Nov. 1809		RH146–50; R157	
	10 Aug. 1812		RH153–5	
	13 Aug. 1818		RH171–2	
	14 Apr. 1823		RH177–8	
	19 Apr. 1825		RH178–9	
	8 Dec. 1826		RH183–4	
Thomson, T. (calico printer)	10 May 1834		RH189	
Watson, H.H.	21 Feb. 1837	**351**		
	9 Feb. 1842	**351**		
Watson, T.	12 Nov. 1834	**352**		
Wettenhall, C.N.	1 Sep. 1818	**353**		
Whalley, L.	11 Dec. 1833	**354**		
Whewell, W.	10 Sep. 1831		RH188	
Yates, J.	26 Oct. 1836	**355**		

3. Letters and statements about Dalton

* – statement

From	*To*			
Babbage, C.	Henry, W.C.	7 Feb. 1854		H185–9
Daubeny, C.	Prout, W.	27 Oct. 1831	**355.1**	C253
Davy, H.*		Feb. 1829		H216–7
Davy, J.*		c. 1830		H217–8
Dockray, B.	Henry, W.C.			H166–8
Egerton, F.*		Jun. 1842		H197–8
Faraday, M.	Henry, W.C.	1 Aug. 1853		H132–3
Forbes, J.D.	Miss Forbes	25 Jun. 1831	**355.2**	M116, 10–1
Giles, S.*				H214–6
Grey, C.	Thomson, C.P.	Jun. 1833		H181
Harcourt, V.	Daubeny, C.	Mar. 1832		P231
Henry, W.	Babbage, C.	1829		H174–7; R199–200; S269–72
	Daubeny, C.	30 Aug. 1831		P227–30

❧ Index of Names

See also the Index of Letters on the previous pages. Note: **Bold** numbers refer to the numbered items in the bibliography. Roman numbers refer to pages in the introduction.

Abbatt, Hannah *see* Tipping, Hannah
Abbatt, Robert **155, 156**
Adamson, R.S. **380**
Allan, John **566**
Allen, Joseph, **310, 433–8, 490**
Alley, P.B. **363**
Anthony, H.D. **760**
Arago, Dominic **156.2**
Ardern, L.L. **392, 607**
Armitage, F.P. **681**
Arrhenius, Svante **654**
Avogadro, Amedeo **483, 484, 637**
Axon, H.J. **728, 800**

B., H. **538**
Babbage, Charles **157, 158**
Backhouse, Jonathan **159**
Bailey, R.W. **565**
Baker, I.H. **439**
Ball, W.W.R. **648**
Banks, John **360.2**
Barclay, A. **686**
Barker, E.H. **326**
Barnes, C.L. **379, 566**
Barnes, H.D. **576**
Barr, E.S. **801**
Barton, Thomas **99**
Baxendell, Joseph **371**
Bealey, Adam **546**
Bealey, John **324**
Beard, J.R. **309**
Bedlow, Robert **802**
Begg, Gordon **472.2**
Beketov, A.N. **737**
Benson, Robt. xiv
Benson, William **321, 322**
Benzenberg, J.F. **618**
Berkenheim, B.M. **723**
Bernard, T. **159.2**
Bernatowicz, A.J. **803**
Bernoulli, Jacques **106**
Berry, A.J. **710**
Berthollet, C.L. **660, 832, 894**
Berzelius, J.J. **51, 98, 236, 277, 279, 490, 671, 889**

Bewley, George **160, 161, 161.2**
Bewley, Hannah, *see* Tipping, Hannah
Bewley, Isabella xiv
Bewley, Richard **611**
Bickerstaff, Robert **325**
Biot, J.B. **161.5**
Birley, H.H. **309**
Bishop, P.W. **749**
Blackett, P.M.S. **1037**
Blamires, David **985**
Bolam, C.G. **804**
Booker, Winifred **457.2**
Boorse, H.A. **929**
Bostock, John **46, 390, 614**
Bowmaker, J.K. **853**
Bradley, John **439.2**
Bradley, William **439.5**
Bradshaw's Jnl **1035–6**
Bragg, Lawrence, **1037**
Brewster, David **490, 513, 518**
Brierley, John **771**
Brindley, W.H. **566, 603**
Brittain, J.H. **532**
Brock, W.H. **806, 807, 930**
Brockbank, E.M. **477, 478, 682**
Brockbank, William **391**
Brockedon, William **162, 440, 925**
Brodie, Benjamin **930**
Brooks, Ann **965**
Broughton, Arthur **105.5**
Brown, F. Madox **416, 441, 442**
Brown, J.C. **662**
Brush, S.G. **941, 950**
Buckley, A.B. **635**
Bud, Robert **937**
Bugge, G. **567**
Burkhardt, G.N. **966**
Buss, R.W. **473, 474**
Butler, F.H. **519**
Butler, R.R. **712**
Butler, Stella **967**
Butterworth, L.M.A. **808**

Cadell, Thomas **162.2**
Calvert, M **475.4**

೭ Index of subjects

Note: **Bold** numbers refer to the numbered items in the bibliography. Roman numbers refer to pages in the introduction